LOCKED in TIME

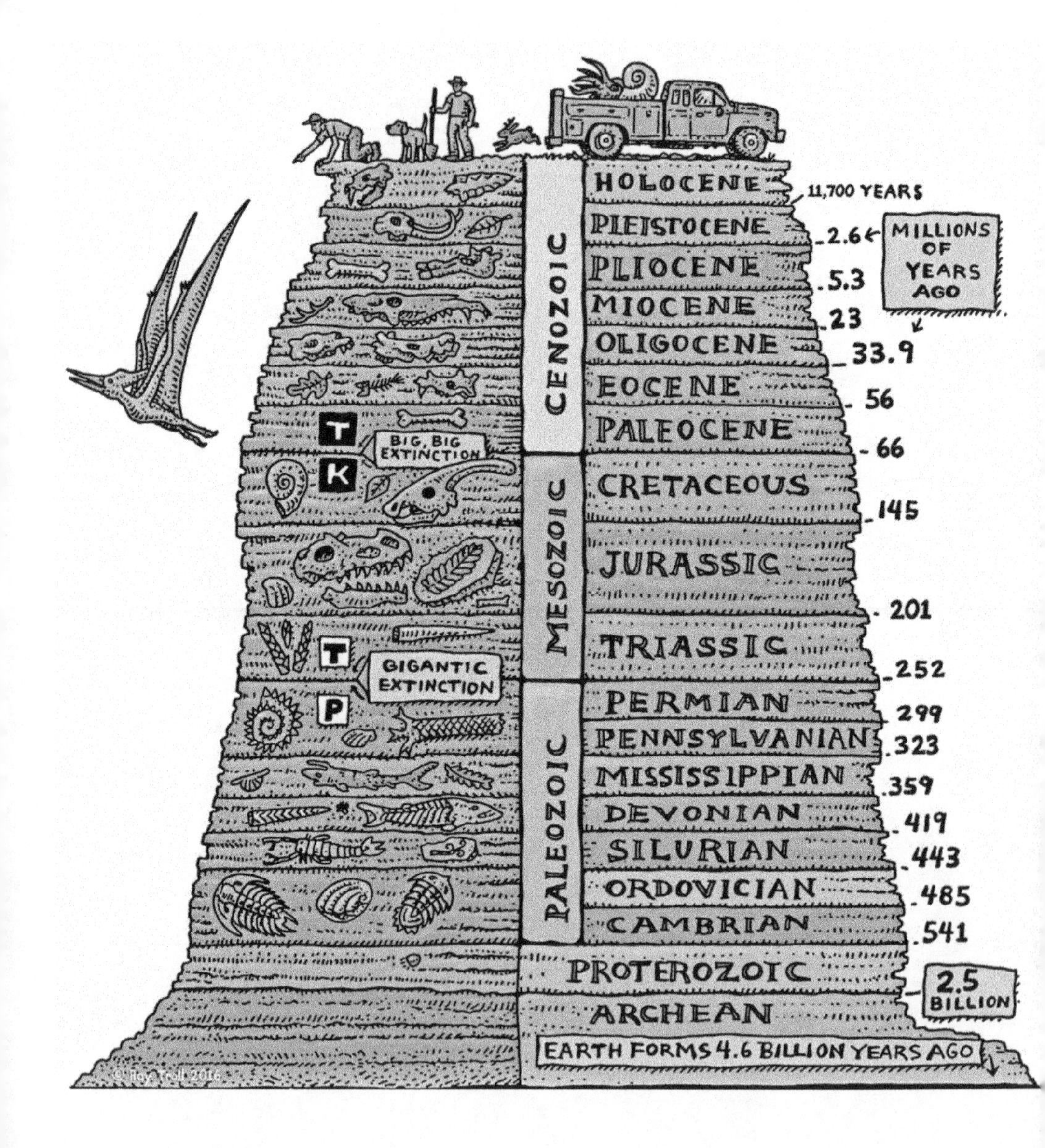
HOLOCENE
11,700 YEARS
PLEISTOCENE
2.6
MILLIONS OF YEARS AGO
PLIOCENE
5.3
MIOCENE
23
OLIGOCENE
33.9
EOCENE
56
PALEOCENE
66
CENOZOIC
T
BIG, BIG EXTINCTION
K
CRETACEOUS
145
JURASSIC
201
MESOZOIC
TRIASSIC
252
T
GIGANTIC EXTINCTION
P
PERMIAN
299
PENNSYLVANIAN
323
MISSISSIPPIAN
359
DEVONIAN
419
SILURIAN
443
ORDOVICIAN
485
CAMBRIAN
541
PALEOZOIC
PROTEROZOIC
2.5 BILLION
ARCHEAN
EARTH FORMS 4.6 BILLION YEARS AGO
© Roy Troll 2016

LOCKED in TIME

Animal Behavior Unearthed in
50 Extraordinary Fossils

DEAN R. LOMAX

Illustrated by Bob Nicholls

Columbia University Press
New York

Columbia University Press
Publishers Since 1893
New York Chichester, West Sussex
cup.columbia.edu

Library of Congress Cataloging-in-Publication Data
Names: Lomax, Dean R., author. | Nicholls, Bob, 1975– (Paleoartist), illustrator.
Title: Locked in time : animal behavior unearthed in 50 extraordinary fossils / Dean R. Lomax ; illustrated by Bob Nicholls.
Description: New York : Columbia University Press, [2021] | Includes bibliographical references and index.
Identifiers: LCCN 2020046670 (print) | LCCN 2020046671 (ebook) | ISBN 9780231197281 (hardback) | ISBN 9780231197298 (pbk.) | ISBN 9780231552080 (ebook)
Subjects: LCSH: Animals, Fossil. | Animals, Fossil–Behavior. | Animal behavior–Evolution.
Classification: LCC QE761 .L66 2021 (print) | LCC QE761 (ebook) | DDC 560–dc23
LC record available at https://lccn.loc.gov/2020046670
LC ebook record available at https://lccn.loc.gov/2020046671

Cover design: Milenda Nan Ok Lee
Cover image: Bob Nicholls
Frontispiece: Ray Troll

For my wonderful mum, Anne Lomax.

Thank you for supporting me every step of the way.

Contents

INTRODUCTION
Unlocking the Prehistoric World

Do you remember the first time you saw a dinosaur skeleton? With your head tilted back, eyes wide open, and experiencing that overwhelming, uncontrollable "wow" factor? For many of us, coming face to face with an intimidating colossus of bones and teeth, an animal unlike anything alive today, would have filled you with an incredible sense of wonder and excitement.

But have you ever pondered what dinosaurs and other prehistoric animals did in their daily lives? Maybe what they ate or if they got sick? Perhaps how the animals cared for their babies and whether they laid eggs or gave birth to live young? Finding answers to these apparently simple questions is an almighty, complex challenge for paleontologists studying fossils. Some fossils, however, are so remarkable that they record direct evidence of such behaviors, capturing specific moments in the lives of long-extinct species. These are among the most fascinating, awe-inspiring, and exceptional of all fossils ever found.

Rather than just preserving an impression or the remains of an organism, these rarest of all fossils do so much more. They give us a detailed glimpse into how prehistoric animals actually *lived*. Although somewhat cryptic, such discoveries provide paleontologists with all (or most) of the pieces of the puzzle to reliably interpret what happened to an animal during its life or at the time of its death.

Take, for example, the fact that in practically every dinosaur film or TV show ever made, the climax will inevitably be a prehistoric battle. In reality, only one example of a pair of dinosaurs fighting to the death has ever been described in the scientific literature. (Another pair of apparently "dueling dinosaurs" has been found but has yet to be formally studied.) This does not mean that fighting was rare in dinosaurs or other prehistoric animals. Rather, it highlights the extreme improbability that such behaviors would be captured in fossils. Consider how slim the chances are that two dinosaurs, dying mid-fight, would be trapped and buried together so they would not just stay intact but then become preserved for eternity, capturing their fight for all time. That paleontologists would then come along millions of years later and discover them still locked in combat is extremely fortunate, especially considering that the fossil could easily have been eroded away and lost forever.

Even though dinosaurs make up only a tiny fraction of what we know about prehistoric life, for many they are the gateway into paleontology and often science in general. In this book, I have included amazing dinosaur fossils that tell unique stories—such as dinosaur parents caring for young and dinosaurs dealing with disease. I have also taken advantage of exquisite fossils from other animal groups whose tales extend hundreds of millions of years into the past. You will be given a glimpse of the mating rituals of creatures preserved forever mid-sex, uncover a giant shark nursery, capture the magical moment of a seagoing reptile giving birth, and discover many more extraordinary fossils.

The fifty stories I have curated in this book will take you on a global journey through the depths of time. Bringing these exciting real events—and the animals themselves—to life, Bob Nicholls has created some beautiful illustrations, one for each of the fifty fossils. All of the illustrations are based entirely on the story contained in each fossil. Most important, they are all scientifically accurate, derived from evidence and scientific research as opposed to simple assumptions.

I want to show that fossils are not just inanimate objects. They are time capsules that provide us with an insight into a vanished world—a world that is unrecognizable and yet so familiar all at once. These fossils offer much more than facts and figures; they showcase that behaviors typical of living animals in the present day have their evolutionary origins nestled far back in time. Through the following stories and illustrations, we can see snapshots of moments in time and learn to appreciate that these were once living, breathing animals that were as real as you and me. This is their story, of lives *Locked in Time.*

DITCHING THE "DUSTY OLD FOSSILS"

Paleontology is going through a golden age of discovery right now. New finds are coming in thick and fast. Whether it is the latest feathered dinosaur with color from Asia, the DNA study of a 12,000-year-old mammal from South America, or the identification of an extinct species of early humans from Africa, the latest and greatest finds pique the interest of many. In today's world, where access to knowledge is at our fingertips, paleontology has never been so openly available.

It is through these types of discoveries and their reports in the media that I became so fascinated with fossils. I was one of those children who absolutely loved dinosaurs and all things prehistoric. You know the type—that annoying kid telling everybody dinosaur facts. Today, I get to do this as my job.

My passion for this project stems from what became a career-defining trip to Wyoming in 2008. This was the first time for me traveling from the UK to the United States; it was also the first time I traveled abroad alone. I was eighteen years old and had sold my *Star Wars* collection (yes, really!) to fund my first professional excavation and research trip to the Wyoming Dinosaur Center, a wonderful museum in the mythical-sounding town of Thermopolis. On my very first day, I was given a tour of the museum displays, and one incredible fossil caught my eye: a large slab of limestone rock that had tiny footprints tiptoeing across its surface. I followed them and unknowingly found myself retracing the footsteps of a 150-million-year-old death scene.

Lying at the end of this extensive track was the culprit, a juvenile Jurassic horseshoe crab. This little arthropod had been flung into a toxic lagoon, landed on its back, and then marched to its death, where it eventually suffocated due to a lack of oxygen. The very fact that this entire moment from millions of years ago was preserved here for all eternity astonished me. It had such a profound impact that it changed the way I thought about fossils.

Fueling my excitement further, I learned that this fossil had not been studied, so I jumped at the opportunity to investigate it and teamed up with one of the then museum paleontologists, Chris Racay. Together we researched the specimen, and our findings were formally described in a peer-reviewed journal a few years later. I was young and eager to learn, and it really was a case of being in the right place at the right time, something that could not be said for the horseshoe crab.

Ever since seeing the specimen, I became fascinated by fossils that had their own unique story of behaviors. It was this very fossil, however, that gave me the idea for *Locked in Time*, a book showcasing the astounding and exciting stories that prehistoric animals can tell.

UNDERSTANDING BEHAVIOR IN FOSSILS

At times action packed and filled with drama, animal behavior can be the most exciting and sometimes rather peculiar aspect of nature. Whether it's an eagle swooping down and snatching its fish prey, a horned lizard squirting blood from its eyes in self-defense, or lemurs using millipede secretions as possible medicine—a hugely diverse and often complex world of behaviors has evolved throughout the animal kingdom.

One of the most frustrating things about studying fossils is knowing that you will never see a living member of that extinct species. It is a strange feeling and one I compare with astronomers and stargazers, who can see and study planets and stars but know they will never be able to visit them. Understanding and inferring behavior from fossils is one of the most daunting and challenging tasks that face paleontologists, although it is one that excites us the most.

Having direct evidence, such as an animal with its last meal preserved or a tooth embedded in a bone, is the only way to know for certain that a prehistoric species engaged in a specific type of behavior. This can be interpreted further through careful, detailed comparisons with living relatives and/or analogues. In some cases, analogues are more pertinent even when there are living relatives of the extinct species. For example, a robin is a living dinosaur, but this does not mean that it would be the perfect analogue for a *Diplodocus*, because they have radically different body plans and modes of life, just as nobody would compare the behaviors of a squirrel with those of a blue whale simply because they are both mammals. In modern ecosystems, we can study animal behavior (*ethology*) directly in real time, which provides a necessary platform for paleontologists to interpret behavior in fossils (*paleoethology*) and can, in turn, reveal important information about ancient interactions between species and their environment (*paleoecology*).

The identifications and interpretations of fossils and their behaviors in this book are based on detailed investigations by paleontologists; in some cases, I have also examined or studied the specimens. However, as is the wonderful nature of science, new evidence might arise from the discovery of additional fossils and further studies, which might offer alternative explanations and change our understanding of the behaviors inferred from these fossils.

I had great fun combing through the extensive literature searching for exceptional fossils with evidence of behavior. There is a lot of research dedicated to fossil behavior, particularly in amber, a specialized mode of fossilization that is famous for capturing behavioral interactions. For these reasons, and to strike a fine balance in the fifty fossils and their stories, I was careful to select a diverse range of fossils illustrating various types of simple and more complex behaviors. Many of these specimens are one of a kind, whereas others are represented by multiple examples and highlight a behavior that was commonly preserved.

As you read this book, I want you to experience the same thrill I did when I saw that 150-million-year-old horseshoe crab death track for the first time. From the outlandish to the surprisingly familiar, these stories are not fictitious like *Jurassic Park*; they are instead the real-life tales of long-extinct animals that are written in the rocks.

1 Sex

The breeding season marks a special time in the life of the male brown antechinus, a mouse-sized marsupial native to Australia. Having not even reached his first birthday, he and all of his male contemporaries suddenly stop making sperm. Whatever stores they have, the time has come to use them. Triggered by this stimulus, over the next couple of weeks the males become sex-crazed. Competing with other males in hopes of impregnating as many females as possible, they have as much sex as their little bodies allow. They mate for hours, to the point of exhaustion. Eventually, all this energy exertion takes a serious toll on the males. Their stress levels elevate, causing their immune systems to collapse. They begin to lose their fur, become dangerously ill with infection, and even bleed internally. Yet, they do not stop, until they all drop—dead. Every single male dies as a result of having too much sex. (That is definitely one way to go out with a bang.) Known as *suicidal reproduction*, or, more technically, *semelparity*, these mini mammals pay the ultimate price for their chance to reproduce.

That evolution would result in such unusual and dramatic reproductive behaviors is a testament to the diversity and complexity of life on Earth. After all, every single living organism on this planet is here as a direct consequence of reproduction; it is a fundamental component of life. Simply, if a species does not reproduce, it will become extinct. Nothing lives forever, but having the ability to reproduce is sort of like cheating death. Individuals can leave a genetic legacy, passing on their own unique characters to the next generation and inadvertently ensuring the survival of their species.

Organisms reproduce through either asexual or sexual reproduction. The former is reproduction without sex and requires just a single parent who can create multiple genetically identical offspring (i.e., clones), ideal for stationary organisms. It is simple and fast, requires little energy, and is found throughout multiple groups including corals, sponges, plants, and insects, but it rarely occurs in fish or reptiles and never in mammals. Conversely, sexual reproduction, or sex, requires both a male and female to create fertile offspring. This is through either internal fertilization (think "Part A slots into Part B") as in mammals, or external fertilization, such as a female salmon releasing her eggs into the water (spawning), ready for the male to release his sperm and fertilize them. This process is more time consuming, requires two individuals, and can take a lot of energy. Just think how much energy we humans invest in finding a compatible partner. Nonetheless, sex leads to variation and genetically unique offspring that are a combination of their parents. This mix-up of genes results in offspring that are more likely to thrive, survive, and reproduce.

Although most individuals reproduce by one method, many organisms have the ability to reproduce in both ways. Though it is rare among vertebrates, numerous invertebrates, such as earthworms and snails, have both male and female reproductive organs (they are hermaphrodites), and many are capable of self-fertilization, although typically they reproduce sexually. Some species of fish, such as clownfish, are also hermaphroditic, but they begin life as male and later switch to female, enabling them to reproduce sexually.

In living animals, it can be easy to determine the sex of an individual, but it is not always about the soft parts. Many have external features that distinguish between males and females. Known as *sexual dimorphism*, sometimes these differences are striking and usually most evident in males. For example, male lions have manes, most species of male deer have antlers, and male peafowl (peacocks) have extravagant, colorful tails (trains). However, occasionally it is the female of the species that displays such features, like phalarope shorebirds, whose females are larger than males and have brighter breeding plumage. In such instances, these features not only distinguish the sex but also play a particular role in display, dominance, and competition. This is driven by sexual selection, a form of Darwin's natural selection, in which such sexual traits might aid in the individual's ability to successfully find a mate and reproduce. Although, when it comes to sex and passing on their genes, each species has its own ways of initiating it, or at least trying. Some of these behaviors, such as dancing to impress or building the best nests and offering gifts, can be elaborate, highly complex, or even deadly (like the brown antechinus competing with their sperm). With such an enormous diversity of living organisms, it is not surprising that a whole host of reproductive strategies have evolved, each playing an important role in the survival of a species.

Well, then, what about fossils? *Fossil* and *sex* are two words that are rarely used together, but why? Without hundreds of millions of years of successful reproduction, you would not be reading this book right now. Although it might seem odd, or perhaps obvious, when you think about it, you and every living thing on this planet are a by-product of prehistoric reproduction, a result of long-extinct ancestors passing on their DNA. Millions of individuals had to survive birth, disease, predation, and natural disasters to reach sexual maturity, find a mate, and reproduce. Generation after generation, species after new species, their DNA flows through your veins. Through the plants in the woods or the birds flying above, the evolutionary origins of each organism are nestled back in time.

Yet, how and what can we actually learn from fossils about reproduction? To quote a certain dinosaur novel that led to a blockbuster movie, "Does anyone go out and, uh, lift up the dinosaurs' skirts to have a look? I mean, how does one determine the sex of a dinosaur, anyway?" This is a good point. Typically, fossils are just the hard parts (such as bones and teeth), and those that do preserve the soft parts rarely provide any indication of sex or information about reproductive strategies. Thus, are we really capable of providing anything more than an educated guess?

It is a fair question, especially as it would appear to be impossible for prehistoric animals to leave behind evidence of reproductive behaviors. We can, however, make use of *functional morphology*—the study of the relationship between the structures of an organism and how its many parts function. Studying form and function has the potential to help answer questions about prehistoric organisms and their lives. There are obvious limitations, but we can use living members of the same animal group (if examples still exist) as analogues for how an extinct organism might have reproduced.

What about when we *do* have evidence of reproduction? This chapter helps to shed light on important steps in the evolution of reproduction and associated behaviors—from organisms preserved in the act of doing the deed to those that show evidence of pregnancy— capturing some surprisingly special, intimate moments from deep in the past.

MOTHER FISH AND SEX AS WE KNOW IT

We humans share a common ancestor with all four-limbed vertebrates, including those that have secondarily lost their limbs (such as snakes), going back more than 400 million years. That common ancestor was a fish, and in an evolutionary sense, we *are* fish. Summarized beautifully in Neil Shubin's *Your Inner Fish*, the story of how we, and all tetrapods, have evolved over millions of years is one of the most extraordinary journeys of evolution. Yet, little direct evidence was known about the reproductive behaviors of these early fish. That, however, changed in 2005 with the discovery of one of the most extraordinary fossil finds of all time.

On the hunt for fossil fish, renowned expert John Long had organized a fossil fishing trip to the Gogo area in the Kimberley region of Western Australia, an area he knew well. About 380 million years ago, in the Devonian period, this area was a shallow sea and represents Australia's first Great Barrier Reef. Reefs today are home to a wonderful array of organisms, and that holds true for this ancient reef. It is known for its exceptional fossil preservation, particularly three-dimensional fish. These fossils are contained inside "Gogo nodules," rounded limestone concretions that, once cracked open—if you're lucky—reveal a fossil inside.

One of the luckiest days would be July 7, 2005. Lindsay Hatcher, a friend of Long's who accompanied him on this trip, picked up a nodule, gave it the old routine crack with a hammer, and struck gold. Showing it to John, he identified it as being that of an extinct armor-plated fish called a placoderm. But he was unable to provide a more specific identification, as the surrounding rock had buried most of the fish. So, it was carefully wrapped up and destined for the lab at the Western Australian Museum in Perth, where the rock could be removed. At the time, neither of the pair had any clue of what great discovery they had made, as the secret of this armored fish was still lying in wait.

Placoderms from the Gogo area are well known. Diverse in shape and size, placoderm fossils have been found at numerous locations around the world, and more than three hundred species have been discovered. They were among the very first fish with jaws, but they also had paired appendages (fins) and some of them developed teeth and, as a result, play an important part in helping us to understand the evolution of modern vertebrates. Studies show that placoderms might have been one of the great ancestors on the same evolutionary line that led to humans, or an evolutionary side branch related to our fishy ancestors.

Having waited in line for more than two years, in November 2007, Hatcher's fossil was finally ready to be cleaned. Due to the delicate nature of the bones, the surrounding rock was carefully removed with a weak solution of acetic acid (strong vinegar). The acid eats away at the limestone rock but does not harm the bones unless left for too long. To help speed the process of rock removal, which can take several months, the specimen was also placed into an acid bath. The preparation revealed an almost complete three-dimensional skeleton, around 15 centimeters in length, preserved with a head and even the braincase. Realizing its significance, and to help describe the fossil, John enlisted the aid of Kate Trinajstic, another world-renowned expert on fossil fish and early jawed vertebrates. Not only was this specimen well preserved, but the team realized that it was an entirely new genus and species; even so, the most sensational discovery had not yet been uncovered.

Some bits of rock still concealed parts of the fish, so although it was potentially risky, they decided to place the fossil back into an acid bath for another dip. The now freshly exposed portion of fish revealed something unusual—a tiny fish skeleton, perfectly matching the adult in anatomical details. Realization hit home. They had discovered the world's oldest pregnant fossil vertebrate, predating the previous oldest by more than 130 million years. Further examination of the little embryo showed that a ropelike structure was twisted around it and connected to the adult female. With the aid of a powerful electron microscope, it became clear that this structure was the giver of life, an umbilical cord—a 380-million-year-old umbilical cord, the first maternal feeding structure preserved in any fossil. Moreover, it was positioned near to what might even be the yolk sac.

This great-grandmother fish was named *Materpiscis attenboroughi*, the former meaning "mother fish" in Latin and the latter after Sir David Attenborough, who drew attention to the Gogo fish site back in 1979, as part of the TV series *Life on Earth*. After identifying the pregnant mother, and eager to find more, the team turned their attention to previously collected fossils from Gogo. To their surprise, they found additional pregnant specimens. However, these specimens belong to other placoderm species, one of which contained three small embryos, positioned in the same region as the tiny skeleton in *Materpiscis*.

Following in the footsteps of these discoveries, in 2020 a separate group of researchers identified yet another pregnant placoderm called *Watsonosteus fletti*. This time the fossil was collected from slightly earlier 385-million-year-old rocks in Scotland. Based on the age, the new fish find snatched the title from *Materpiscis* as the world's oldest known pregnant fossil vertebrate.

The fact that these fish are carrying embryos is conclusive evidence that they were viviparous (gave birth to live young). And although we have no idea how long their gestation period was, one thing for certain is that this was the result of internal fertilization; there was no egg laying here. That means these placoderms were having sex. Long before the birds or the bees had even evolved, these armored fish had perfected sexual intercourse—sex as we know it. But how exactly were they doing it? To answer that, we have to look at some of the *hard* parts of the skeleton.

Most living fish that reproduce by internal fertilization do so with the aid of sex structures formed from their pelvic or anal fins. For example, male and female chondrichthyans (sharks, etc.) can be distinguished because males possess penis-like claspers that they use during copulation. Such sexually dimorphic structures have been found among several placoderm species, distinguishing the males from females. The earliest known examples, and thus the oldest joint evidence we have of sexual intercourse, are around 385 million years old and belong to a thumb-sized species called *Microbrachius dicki*. The "*dicki*" is just an amusing coincidence, as they were named after the discoverer of the first fossils, Robert Dick. Lots of specimens have been found, including some from Estonia and China, but most are from Scotland. They show that the males have huge and distinct hook-shaped bony dermal claspers, whereas the females have paired blade-like genital plates. A specimen from Estonia even shows one male clasper attached to the female plate. During sexual intercourse, as these animals swam side by side, probably with their bony arms interlocked, the male would have inserted the tip of one of his large claspers into the female's cloaca, where the genital plates held it in place and aided delivery of the all-important sperm.

It had always been assumed that external fertilization was the primitive mode of reproduction for fish like these placoderms. However, these remarkable discoveries, of preserved bony sex organs and even pregnancy, shows that these early fish were among the first vertebrates to copulate and bear live young.

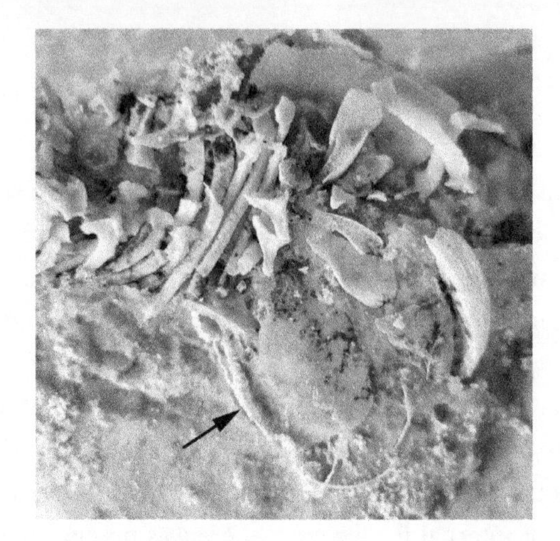

FIGURE 1.1 Close-up of the pregnant mother fish, *Materpiscis attenboroughi*, with fragmentary remains of the embryo and ropelike umbilical cord (*identified by the arrow*).

(Courtesy of John Long, Flinders University)

FIGURE 1.2. Male (*left*) and female (*right*) *Microbrachius dicki*. The male is identified by the presence of penis-like claspers and the female has paired genital plates. (A) arm; (C) clasper; (G) genital plate; (H) head.

(Courtesy of John Long, Flinders University)

FIGURE 1.3. (Opposite). *Love entwined.*

A pair of mating *Microbrachius* take center stage, with the male on the right and female on the left, while another is giving birth in the background.

BOB NICHOLLS 2020

DINOSAUR SEX DANCE

Birds are dinosaurs. From a hummingbird to a turkey and from a pelican to an emu, every single species of the more than 10,000 living birds are dinosaurs, a group of theropod dinosaurs to be exact. Among many features, similarities in their skeletal anatomy, the presence of feathers, and even direct behaviors such as brooding link birds and theropods. And now we have evidence that suggests that some theropod dinosaurs "danced" to impress the opposite sex.

This dancing is, in fact, a behavior called *lekking*, which many ground-nesting birds do today. During or before the breeding season, typically gathering in groups, males put on an impressive show to attract the attention of females. Females watch as the males compete with one another through intense vocalization and silky dance moves and by showing off their feathers. However, one important act that males also tend to display is their nest-making abilities, which they demonstrate by scraping the sediment with their feet, signaling to the females that they are strong and capable of building a successful nest.

Large fossil scrape marks, up to 2 meters in diameter and dating from 100 million years to the Cretaceous, have been found at four sites in Colorado. The scrape marks are not simply a one-off discovery but are known from multiple specimens. One site revealed more than sixty distinct scrapes in a single area of up to fifty meters long and 15 meters wide. Each of the sites is well known for dinosaur tracks, but it was only in 2016 that these fossil scrapes were recognized as such and formally described. Some of the scrapes were found at the famous site known as Dinosaur Ridge (or "Dinosaur Freeway"), a National Natural Landmark, where dinosaur tracks were first discovered in 1937.

The scrape marks differ in their size, depth, and distribution, but most consist of parallel double troughs resulting from multiple claw scrapes made by both feet. Some of these scrapes show complete outlines of the three-toed theropod footprints. Based on the length of the various footprints, it can be estimated that the creators of these marks had total body lengths of between 2.5 and 5 meters. It is unfortunate that the bones of the theropod that made these scrapes were not found at the site, but body fossils from this particular rock unit (known as the Dakota Sandstone) are exceptionally rare.

FIGURE 1.4. Several prominent scrape marks made by a large theropod at Roubideau Creek, Delta County, Colorado. Researchers Ken Cart (*lower left*) and Jason Martin (*upper right*) for scale.

(Courtesy of Martin Lockley)

FIGURE 1.5. The author (*left*) and paleontologist Lou Taylor (*right*) inspect some very large scrapes at the famous Dinosaur Ridge site, Morrison, Colorado.

(Courtesy of Malcolm Bedell Jr.)

Bob Nicholls 2020

A dancing dinosaur conjures up a vivid image, and it might seem fair to question whether these scrape marks are really the result of a dinosaur courtship display. This is the question that the research team, led by dinosaur footprint aficionado Martin Lockley, had to address.

A few other interpretations were considered during examination of the traces, notably whether the four sites represented actual nests or colonies or were evidence of dinosaurs digging for food, water, or perhaps shelter. Yet, there is no evidence of eggs, eggshells, or any indicator of a well-defined nest typical of known dinosaur nesting sites. The geology indicates that during this time, the environment was very wet, and so it seems that water would have been readily available. It is plausible that the theropods might have dug for food, perhaps smaller animals underfoot, but there is no evidence of burrows or any indication to suggest that a carcass was buried at any of the sites.

These Cretaceous sites are, instead, the remnants of a mass courtship display of multiple male theropods that were probably seasonally active and gathered to strut their nest-building abilities to woo onlooking females in a competitive sexual display. Once a mate had been selected, just like in living birds, the nest would have been made close by; however, the breeding or nesting sites have not yet been found and might never have been preserved. Such scrape-marking behavior by a theropod dinosaur is the first evidence directly linking dinosaur mating behavior with that seen in living birds. We can only wonder what sounds and spectacular displays large theropods made during such courtship ceremonies.

FIGURE 1.6. (Overleaf). *The lady looks on.*

Multiple large theropods engage in a fierce dino dance competition in hope of wooing a female who is watching the show. We do not know which species created the scrapes, but *Acrocanthosaurus* or something similar (pictured is a speculative species) is a good candidate.

LIFE IN DEATH: LIVE BIRTH IN PREGNANT "FISH-LIZARDS"

For hundreds of millions of years, fish were the only vertebrates in the sea. But long after the armored fish and way before great white sharks or killer whales, a group of terrestrial reptiles entered the watery realms for the first time, eventually replacing fish as the ocean's top predators. They dominated for more than 180 million years, from the beginning of the Triassic through the end of the Cretaceous. Marine reptiles had to overcome a variety of obstacles in the water but none more important than giving birth and ensuring the survival of the species. The ichthyosaurs (Greek for "fish-lizards") are among the most famous of their kind and were the first fossilized marine reptiles to shed light on their mode of reproduction.

Ichthyosaurs are in no way, shape, or form swimming dinosaurs, although they have been glorified by the media as such. Unlike dinosaurs, among many other features, ichthyosaurs had fin-like limbs and lived exclusively in the water. They are a truly fascinating group of reptiles that appeared in the fossil record almost 20 million years before the dinosaurs. They were among the first large extinct reptiles brought to the attention of science, mainly through the efforts of Mary Anning of Lyme Regis, Dorset, a late Georgian to early Victorian paleontologist who collected hundreds to thousands of specimens. Ichthyosaurs were named at least twenty years before the word *dinosaur* was even invented. Based on their skeletal anatomy, we know that they evolved from an early land-dwelling ancestor, although what that animal was is still to be determined. In life, the typical ichthyosaur, with its long snout and streamlined body, looked like a primeval cross between a shark and a dolphin. Some of the early forms were small and primitive, akin to lizards with flippers, whereas others approached the size and shape of blue whales.

Because ichthyosaurs are reptiles and have fins that superficially resemble those of sea turtles, it was assumed that they, too, must have come ashore to lay their eggs. An issue with this hypothesis is that they had evolved highly specialized body shapes so well adapted to a marine environment that it seemed impossible for them to leave the sea. So, did they lay eggs in the water like amphibians or give birth to live young? Considering that ichthyosaurs were the first large-bodied reptiles in the ocean and were successful in reproducing, with tens of thousands of fossils found globally, unlocking the mystery behind their reproductive strategies can yield crucial information as to how early reptiles adapted to life in the water.

Our initial clue to this mystery was discovered more than two hundred years ago when the first of many ichthyosaurs was found with small skeletons trapped between its ribcage. Most of these finds came from several Jurassic quarries near the town of Holzmaden, Germany. Their association seemed to show pretty compelling evidence that these animals fed on members of their own species—they were cannibals. This view was generally accepted, although, as more ichthyosaurs were found with multiple small skeletons inside their ribcages—some even containing more than ten individuals—evidence began to accumulate to counter this idea.

A major discovery was made in 1846, when Joseph Chaning Pearce, a prolific collector who amassed one of the largest fossil collections in the United Kingdom, studied an *Ichthyosaurus* skeleton that he had collected from a small village in Somerset. As part of my research, I have had the privilege of studying this same specimen, which is on display today at the Natural History Museum in London. He identified a tiny, almost complete skeleton preserved at the end of the ribcage, positioned in the pelvis region. The position of this tiny skeleton was too far removed from the stomach area and therefore could not possibly be the animal's last meal—a pivotal piece of evidence for live birth in ichthyosaurs. If correct, this would be the first record of a pregnant fossil reptile.

In one study after the next, the debate raged until the 1990s, when two major studies concluded that these ichthyosaurs were falsely accused of cannibalism. Rather, virtually all of them were, indeed, pregnant. Lending support for this interpretation is evidence that the bones of the small skeletons are not etched by stomach acids and do not show any sign of bite marks, and it is highly unlikely that a larger individual would eat multiple babies at the same time. This evidence meant that the pregnancy interpretation was starting to make perfect sense.

Today, more than one hundred pregnant ichthyosaurs, all belonging to one particular species called *Stenopterygius*, have been discovered in the Holzmaden area. The heads of most of the unborn embryos are pointing toward the head of the mother. Based on this position, it was assumed that—like dolphins and whales—the embryos were typically born tail first, with their nostrils emerging last to help prevent them from drowning during labor. Proving this theory correct, and crushing any remaining doubts that these animals did not give birth to live young, was the incredible discovery of specimens in labor.

One extraordinary Holzmaden ichthyosaur that was pregnant with quadruplets captures the unfortunate moment when one of her embryos became stuck in the birth canal with only its head still remaining inside the mother. Sadly, the mother likely died during labor, resulting in the death of not one but five animals. Adding support to this tragic scenario, a specimen of *Chaohusaurus* with triplets was recently collected from eastern China. This discovery is particularly exciting because it is from the Early Triassic period, around 248 million years ago, close to when ichthyosaurs first appeared. In this specimen, however, one embryo is emerging from the mother's pelvis in a head-first birth position, along with a newborn preserved outside of the mother, showing that she had already given birth to at least one offspring. This suggests that the terrestrial ancestors of ichthyosaurs likely bore live young head first, and that early ichthyosaurs also still had the head-first

FIGURE 1.7. The extraordinary pregnant ichthyosaur, *Stenopterygius quadriscissus*, caught in the act of giving birth. One of her would-be newborns is stuck in the birth canal, and three embryos remain inside her ribcage.

(Courtesy of Cindy Howells)

BOB NICHOLLS 2020

delivery mode, which later changed to tail first. This particular specimen represents the oldest vertebrate fossil in the world to capture the actual moment of live birth.

Even though these events happened very long ago, this moment in time has given us direct evidence of reproductive biology in these long-extinct reptiles, something that could not have been figured out were it not for such stunning discoveries.

Their ability to bear live young is likely to be one of the major reasons that ichthyosaurs were so successful. In such circumstances, live birth has several advantages, most notably that the young are (by and large) born highly developed and able to feed and fend for themselves almost immediately. For example, baby ichthyosaurs, or what I like to call "icklets," arrived fully equipped, with cone-shaped needle-like teeth ready to feast on fish.

Live birth does have its limitations, though, especially with regard to how it takes a toll on the mother, who must catch and eat enough food that her embryos are sufficiently well nourished and therefore well developed at the time of birth. One of the biggest concerns would have been a lengthy underwater labor. Like whales and dolphins, these animals needed to breathe air, which meant that if the labor was particularly long, or if there were additional complications, the mother would have had to return to the surface to breathe. She would then run the risk of attack from a predator or, worse, if left too long under water, death by drowning. Immediately after birth, the newborn icklets also would have needed to swim to the surface to take their first breaths and run the same risks.

With these potential issues, live birth at sea is rife with risk, but it was successful for extinct marine reptiles and continues to be so for marine mammals. As for the ichthyosaurs, they were among the first large and secondarily aquatic animals to accomplish this remarkable feat of evolutionary adaptation, enabling them to fully commit to a life at sea.

FIGURE 1.8. (Opposite). *Mortality in the blue.*

A pregnant *Stenopterygius* having complications during labor in the warm Jurassic sea roughly 180 million years ago.

JURASSIC SEX: CAPTURED FOREVER IN THE ACT

Sex is happening right now. Every second of every day, in the sea, on the land, and in the sky, animals are having sex, passing on their genes to the next generation. But imagine being caught in the act for millions of years, only for scientists to study you in every detail and share their findings with the world, for all to see. You must be the unluckiest of individuals to become fossilized mid-sex.

Fossils of animals caught "doing it" would seem almost impossible, and as you might expect, the record of mating fossils is exceedingly rare. Consequently, only a couple of vertebrates have been found mating, and the vast majority of mating fossils are invertebrates, represented by around fifty specimens. Most are insects trapped in amber, including moths, mosquitoes, bees, and ants, caught off guard by sticky resin seeping from trees. But the earliest record of a mating fossil is a pair of froghoppers delicately preserved in a 165-million-year-old Jurassic rock.

Named for their froglike face and record-breaking jumping (hopping) abilities, froghoppers are a group of tiny sap-sucking insects represented by approximately three thousand living species around the world. Immature froghoppers (nymphs), called "spittlebugs," produce a foamy, spit-like froth on plants, which acts as a sort of bubbly cocoon and protects them as they feed on plant sap.

The fossilized pair of mating froghoppers was collected from Daohugou Village in Inner Mongolia, northeastern China. Here, the Middle Jurassic rocks have yielded many extremely well-preserved fossils, including more than twelve hundred froghoppers, each less than 2 centimeters long. A volcanic eruption led to the fine preservation of the fossils, as ash was deposited into a nearby lake and the animals were swept inside and buried.

These specimens are kept in the vast fossil insect collection at the College of Life Sciences at Capital Normal University in Beijing, a collection that comprises more than 200,000 insect fossils. The froghoppers were found to belong to an extinct family and were identified as a new species, *Anthoscytina perpetua*; the species name being from the Latin *perpet*, meaning "eternal love," in reference to their everlasting embrace.

Although they are lying together face to face, can we be certain that this pair of froghoppers really were in the act of mating? Two hundred of the twelve hundred fossils were distinguished by the presence of well-preserved male and female sex organs. The males have a long, tubelike *aedeagus*, and the females have a saclike *bursa copulatrix*. Not quite a penis and

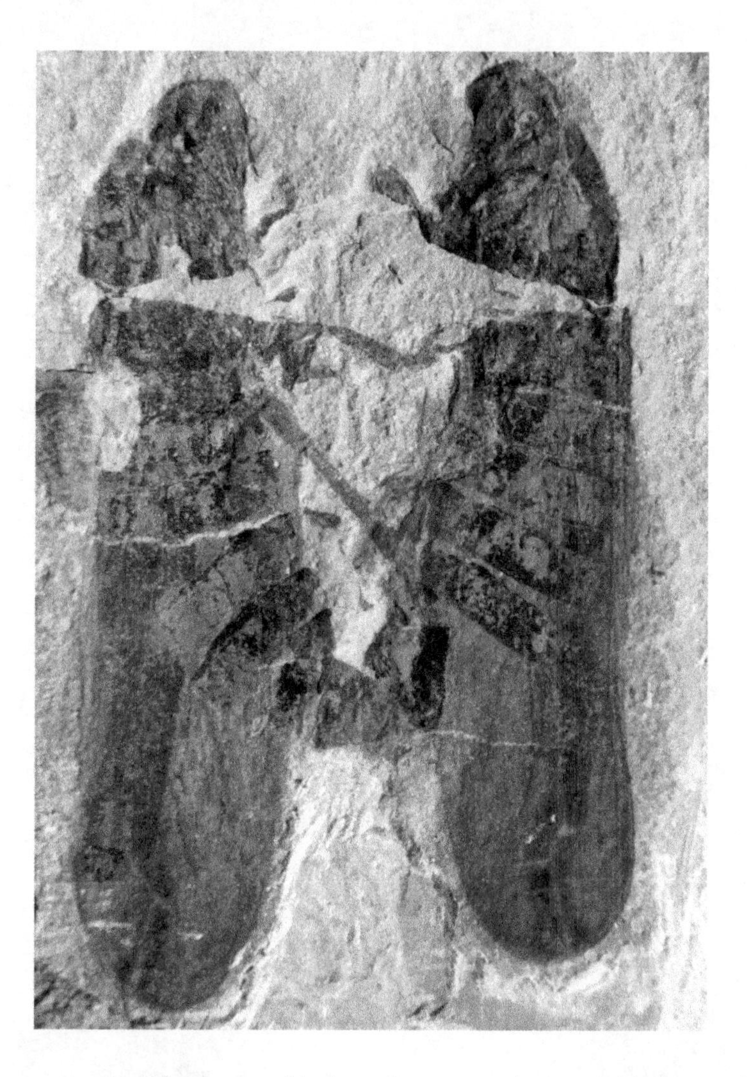

FIGURE 1.9. Fossil froghoppers (*Anthoscytina perpetua*) preserved in the act of mating. The male is on the right and female on the left.

(Courtesy of Dong Ren)

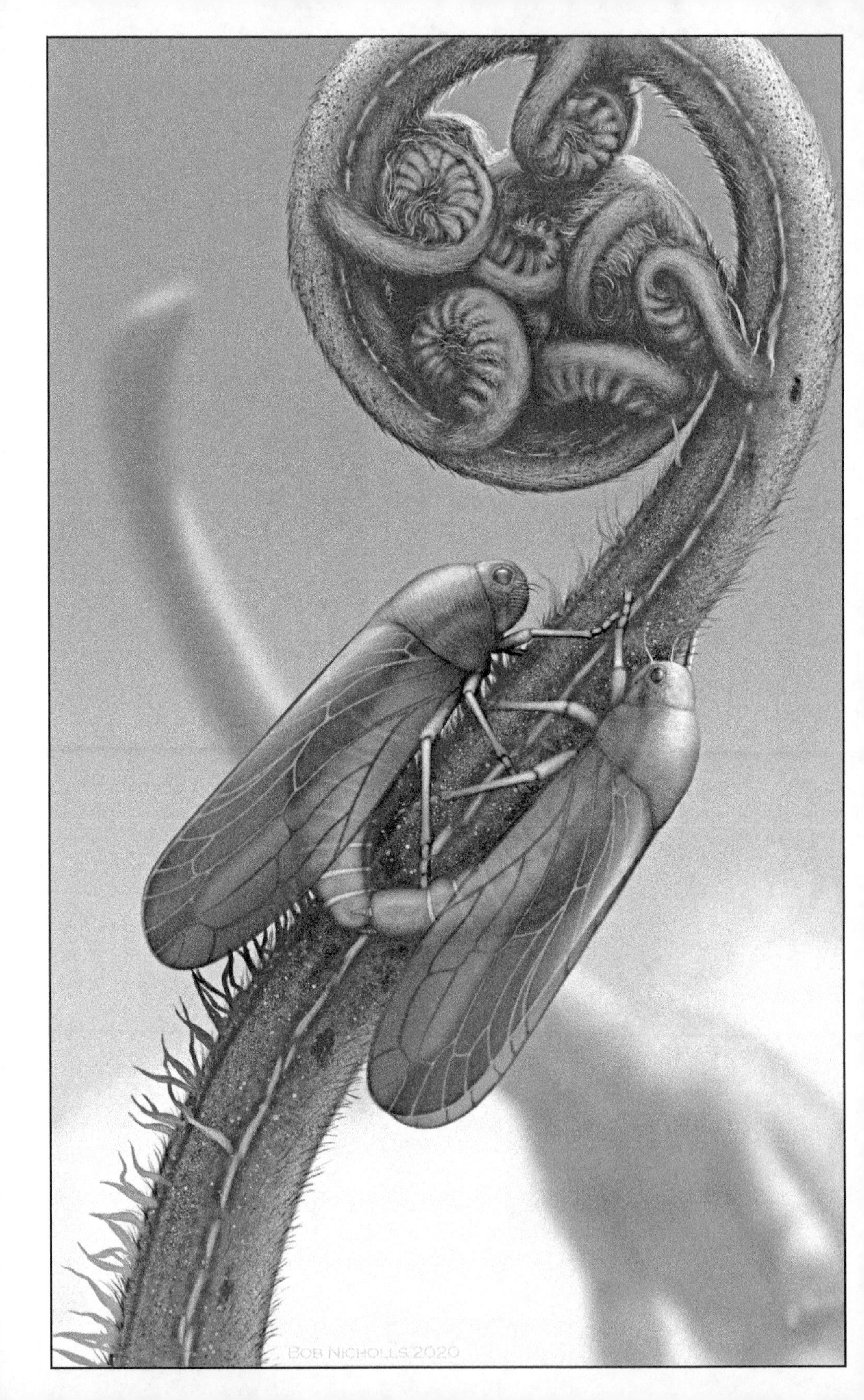
BOB NICHOLLS 2020

vagina as in mammals, but you get the picture. (The same is found in living froghoppers, which also have symmetric genitalia.) The mating pair are lying belly to belly but were likely to originally have been side by side, the typical mating position of modern froghoppers, and the male's aedeagus is inserted directly into the female's bursa copulatrix. Where the aedeagus inserts, the body is curved, and this shows that the male's last few body segments were flexible and could be twisted during sex, which would aid with the insertion of the aedeagus.

There is no denying that this fossil represents a pair of tiny mating froghoppers. Their genitalia and mating position match those seen in their modern-day counterparts and shows that this sexual behavior has remained static for 165 million years.

FIGURE 1.10. (Opposite). *Froghopper sex.*

A pair of *Anthoscytina perpetua* froghoppers engage in a Jurassic sex session as an enormous sauropod passes by, completely oblivious.

A PREGNANT PLESIOSAUR

Plesiosaurs are some of the most iconic of all prehistoric animals. Like the ichthyosaurs, they often have the misfortune of being labeled as swimming dinosaurs but were a carnivorous group of aquatic, mostly marine, reptiles that swam using their four large, winglike paddles or flippers. Since their discovery more than two hundred years ago, scientists and the public alike have pondered how these wondrous animals gave birth. Was it on land or in the water, and did they lay eggs or bear live young? Studies of the mechanics of their skeletons, particularly of their inflexible flippers and how they are weakly attached to the body, implies that plesiosaurs were almost certainly incapable of moving on land and consequently must have given birth to live young in the water. Unlike their ichthyosaur contemporaries, not a single specimen provided any corroborative evidence of their reproductive behaviors. That is, until 2011.

Actually, 2011 is not quite correct; technically, we have to go back to 1987. That year, Charles Bonner, a seasoned fossil hunter, was routinely hiking on his family property at the Bonner Ranch in Logan County, Kansas, and came across a series of fossilized bones that were poking out of the rocks. Thinking it might be something special, Charles, with help from his family, dug up the find. The bones turned out to be that of a 78-million-year-old plesiosaur, which was largely complete and articulated (with most of the skeleton in place with the bones still connected). Not only that, but a smaller skeleton was present near the ribcage. Paleontologists from the Natural History Museum of Los Angeles County received word of the discovery and were extremely excited. Inspired by this reaction, the Bonner family wanted scientists to study their find and kindly donated it to the museum. At the time, paleontologists were quietly confident that this specimen might represent a mother and child preserved together, the first unequivocal evidence that plesiosaurs gave birth to live young. However, the specimen was still buried inside a significant chunk of rock, and this assumption could not be verified until the excess rock was removed.

It might seem odd to read that marine fossils were found in landlocked Kansas, but during the mid- to latter part of the Cretaceous, a warm inland sea known as the Western Interior Seaway stretched across what is now the American Midwest, roughly splitting in half what is today the United States.

Fast-forward to the mid-2000s, when plans were in motion to create a new paleontology gallery at the Natural History Museum of Los Angeles County, including a section dedicated to marine fossils, and attention was turned to the plesiosaur. Now, it was not simply a case of leaving the specimen to languish in the collections for years. Quite the opposite. Often it can cost a lot of money to clean and prepare such delicate and rare fossils, so finances had to be sought. After several years of skilled and careful preparation, the specimen was finally fully exposed and ready to be studied prior to its being placed on display.

The head of the museum's research and collections department, Luis Chiappe, who oversaw the cleaning of the specimen, realized its scientific importance and contacted Robin O'Keefe, a paleontologist at Marshall University in West Virginia. O'Keefe has spent many years studying plesiosaurs and is recognized as a world expert. Together, the duo set about unravelling the story that this specimen contained.

First, the skeleton was compared with known plesiosaur species and identified as a large adult example of the short-necked plesiosaur *Polycotylus latippinus*. This species was first described in 1869 based on a fragmentary specimen, also collected in Kansas. With the species identification nailed down, attention was turned to the all-important possible fetus. This small skeleton comprises an incomplete, poorly ossified, and largely disarticulated mass of bones. Counting all of the individual elements, about 60 to 65 percent of the skeleton is preserved, including numerous vertebrae, ribs, and parts of the pelvis and shoulder girdle, some of which are intermingled with the adult's right fore-paddle. Although this small individual was found with the adult, suggesting that it might be a fetus, other scenarios had to be ruled out, particularly whether the specimen simply came to rest next to the adult after it died or whether the adult had eaten a juvenile.

Countering these interpretations, first of all, parts of the small skeleton are within the body cavity, so there is no way that the little one came to rest at the side of the large adult. The bones of the small skeleton are consistent with *Polycotylus latippinus*, the species identification of the mother. Plus, there is no evidence to suggest that it was consumed by the adult, as there are no signs of tooth damage to the bones or abrasion from stomach acid, as would be expected for a last meal. Undeniably, this represents the skeleton of an unborn fetus.

Compared with the adult individual, which has an estimated total length of 4.7 meters, the bones of the fetal skeleton look very small, but looks can be deceiving. When measured and pieced together, the fetus is large, with

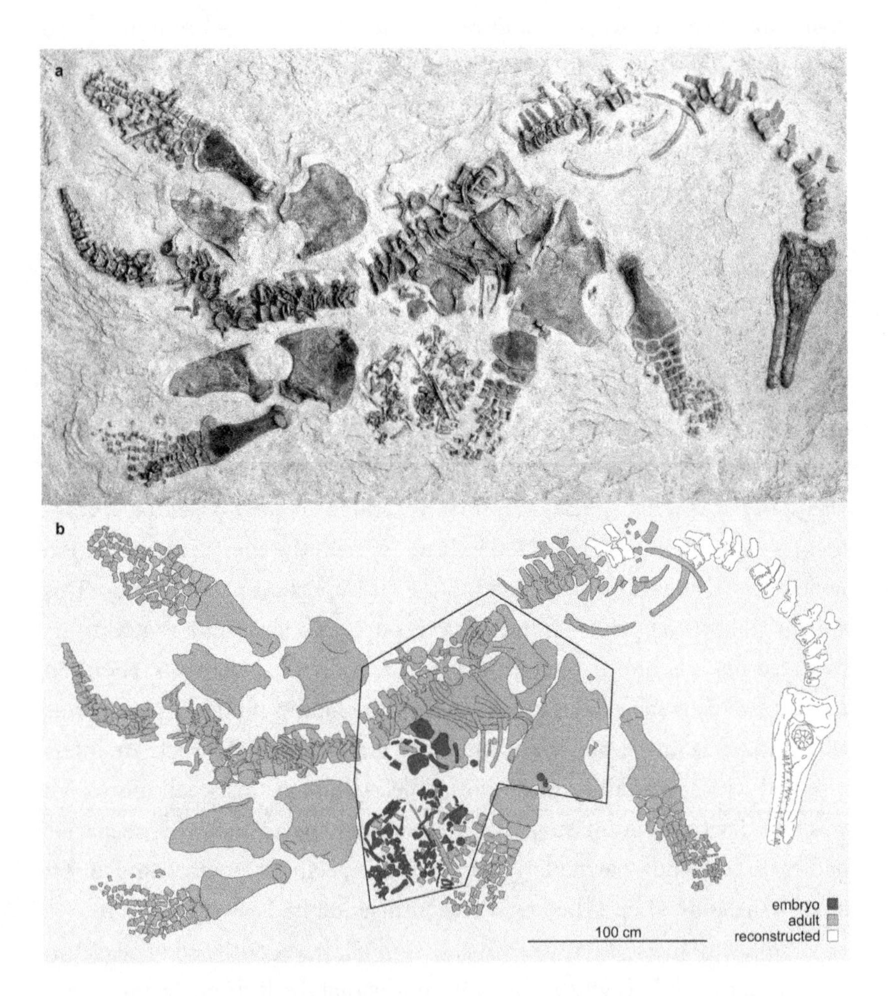

FIGURE 1.11. Photograph (*A*) and interpretive drawing (*B*) of the pregnant plesiosaur, *Polycotylus latippinus*.

(Courtesy of the Dinosaur Institute, Natural History Museum of Los Angeles County)

BOB NICHOLLS 2020

a length estimate of 1.5 meters—32 percent of the mother's length. However, based on the lack of a fully ossified skeleton, the fetus was not full term, and estimates predict that it would have been around 40 percent the length of the mother at birth. That is a big baby. In fact, that is the equivalent of a human giving birth to a six-year-old!

Other marine reptiles, such as ichthyosaurs, typically gave birth to multiple relatively small offspring instead of a single large offspring. This is what is known as an *r-selected reproductive strategy*, whereby having multiple individuals might heighten the chance of survival of a couple of the young, but it is unlikely that all would survive. It is a case of quantity over quality; it also has less of a bearing on the mother, meaning that she invests little or no energy on parental care. By contrast, the presence of a single large fetus preserved inside the mother plesiosaur is what is known as a *K-selected reproductive strategy*, in which all the mother's time and energy is invested into a single baby. This discovery in a plesiosaur is unique among marine reptiles. Of course, this is just one plesiosaur, so it is impossible to say whether all plesiosaur species did the same. Nevertheless, today, this strategy is adopted by marine mammals, which invest their time on a single fetus, some even having a gestation period of up to two years. Marine mammals also show extended parental care and social behavior, which hints at the possibility that plesiosaurs might have cared for their young as well, perhaps in groups.

This spectacular find is the only pregnant plesiosaur recorded so far. Its discovery solves a two-hundred-year-old mystery of their reproductive behaviors and unlocks a wave of fascinating insights into the life cycle of plesiosaurs.

FIGURE 1.12. (Overleaf). *The K strategy.*

A pod of *Polycotylus latippinus* fend off a shark (*Squalicorax*) as one of their members gives birth to an enormous baby.

WHEN WHALES GAVE BIRTH ON LAND

The blue whale is the largest animal on the planet. With a recorded maximum length of 33 meters and weighing more than twenty-five fully grown African elephants, they are perhaps the largest animal ever to have evolved, rivaled only by enormous long-necked sauropod dinosaurs and possibly giant ichthyosaurs. Modern whales are divided into two main groups: the baleen whales, in which the blue whale belongs; and the toothed whales, such as sperm whales as well as dolphins and porpoises. They are collectively known as *cetaceans*. Being mammals, they must breathe air, but they are so highly adapted to an aquatic lifestyle that they are incapable of surviving on land and do everything in the water. These majestic marine mammals might rule the world's oceans today, but their origins lie on the land.

The impact of an enormous asteroid that famously wiped out the non-bird dinosaurs 66 million years ago at the end of the Cretaceous sent shockwaves through the oceans. The great marine reptiles that had reigned supreme for millions of years became extinct. Their disappearance left a void that would eventually be filled by a new group of aquatic animals, the cetaceans. Still, it took more than 20 million years of evolution before they solidified their place as the ocean's top predators.

About the size of a wolf, with well-formed, functional legs and no blowhole, these early walking whales looked unlike any whale we see today. They originated in what is now India and Pakistan around 54 million years ago, near the middle to late Eocene, and are known from multiple species. Their skull and skeletal anatomy, particularly of the ear and ankle bones, supports their identification as ancestral whales. They are deduced as having been amphibious, living in coastal environments, feeding on fish in the sea, and resting, mating, and giving birth on land. You might like to think of their mode of life as being outwardly similar to modern-day marine iguanas, who feed in the sea but mate and lay their eggs on land.

One such amphibious early whale is *Maiacetus inuus* from the Kohlu District in Pakistan, known from two adult skeletons collected in 2000 and 2004 and described in 2009 by renowned fossil whale expert Philip Gingerich and colleagues. It lived approximately 47.5 million years ago and was slightly larger than the earliest whales, at around two and a half meters long, and possessed well-formed, albeit relatively short, legs and feet. Its fingers and toes were particularly long and were almost certainly webbed, indicating that this animal swam with its flipper-like feet but was capable of walking on land. As these ancestral whales are usually known only from poorly preserved or fragmentary remains, this was an impressive find, but a secret about early whale birth was also contained in the fossil.

Preserved between the ribs of one of the *Maiacetus* specimens lies the delicate skull and partial skeleton of a single large fetus. The name *Maiacetus*, after the Greek *Maia* and *ketos*, translates to "mother whale," in reference to the pregnant nature of the fossilized whale. This represents the only fossil fetus of an early walking whale ever discovered. The presence of a single fetus matches that found in whales today, which almost always give birth to just a single calf. Unlike in modern whales, in which the calves are delivered tail first to prevent them from drowning during labor, the fetus in *Maiacetus* is in a head-first birth position—the norm for land mammals—providing further evidence that these early whales gave birth on land.

The skull of the fetus is no longer than 17 centimeters, and, accounting for missing parts, it probably had a total body length of around 65 centimeters, a quarter of the length of its mother. Its large size, coupled with the well-developed skull, which also has defined permanent first molars, indicates that it was probably a near-term fetus. As in living marine mammals, this advanced stage represents precocial development, whereby the newborn would have been mobile, able to walk or swim freely and avoid predators, and potentially catch prey to supplement its mother's milk.

Clearly, the pregnant *Maiacetus* is female. The other adult specimen, however, is the more complete of the two, is slightly larger in overall body size, and has canine teeth that are 20 percent larger than in the female. These minor differences appear to be the result of sexual dimorphism, the male being the larger of the two. This seems plausible considering that similar differences exist between males and females in both terrestrial and marine mammal species today.

FIGURE 1.13. (Opposite). *Headlong into life.*

A walking whale (*Maiacetus inuus*) calf emerges from its mother head first on the beach.

BOB NICHOLLS 2020

The transition from walking whales to fully aquatic species has become a textbook example of evolution. Whales did not necessarily do something new; rather, they exploited a niche that was waiting to be filled. The discovery of this extraordinary pregnant walking whale has illuminated our understanding of how and where (on land) these early amphibious whales gave birth, a process that changed through time to coincide with their adaptation to marine life. Great parallelism can be observed with what was achieved many millions of years previously, when marine reptiles evolved from a live-bearing terrestrial ancestor that eventually made the plunge into the water and never looked back.

SEXING A CRETACEOUS BIRD

With their fancy feathers, spiky crests, and exquisite colors, birds exhibit some of the most visually spectacular display structures in the animal kingdom. Such traits play an important role in sexual selection by enabling birds to ooze sex appeal and stand out from the crowd. By and large, males are usually the most colorful sex and often sport extravagant showy feathers that are used in mating displays. Just think of the luxurious plumage in male birds of paradise, for example. On the other hand, there are plenty of examples of bird species that display subtle differences between sexes, and some are distinguished only by the sounds that they make.

Considering the sexually dimorphic significance of feathers in living birds, it is fair to assume that prehistoric species probably differed in plumage, too. Such differences might be found in fossilized birds that have preserved feathers, but how can you be sure that any variation is an indicator of sex? As troublesome as it might appear, having many examples of exceptionally preserved specimens with feathers can provide answers.

In 1995, a toothless, beaked bird called *Confuciusornis sanctus* was described from the Jehol Biota of Liaoning Province, northeastern China, an area famous for the discovery of many feathered dinosaurs (including birds). It was initially thought to have been collected from Late Jurassic rocks but was later correctly shown to be from the Early Cretaceous period, living around 125 million years ago. Since that first find, several thousand specimens of this crow-sized primitive bird have been unearthed. Many are so exceptional that they contain beautiful feathers and often skin, some even showing scaly skin on the feet and a keratinous sheath on the claws.

Extending across the body from head to tail, feathers in *Confuciusornis* are unmistakable. However, a remarkable difference in the plumage exists among specimens. There are two distinct variations, each with the same skeletal anatomy but differing in the presence or absence of two very long tail feathers. These streamer-like feathers are longer than the entire skeleton and protrude from a very short tail formed by the fusion of tail vertebrae into one solid bone (the *pygostyle*), as in modern birds. Even more amazing, several *Confuciusornis* specimens have been found with both variants preserved in the same slab of rock.

In living species like the male ribbon-tailed astrapia bird of paradise, whose tail feathers are over three times the length of its body, the presence or absence of such ornamental feathers reflects sexual dimorphism. This difference in *Confuciusornis* plumage has led to the interpretation that specimens with long tail feathers were males and those without them were females; something on which most researchers agree. Other explanations have suggested that the variation might be the result of differential molting, whereby birds shed old, worn feathers and replace them with new ones, or that the presence of tail feathers could be related to age, found only in sexually mature individuals.

To test the male-versus-female hypothesis further, one team extracted tiny bone samples from multiple *Confuciusornis* specimens, including three supposed males and six females. Under the microscope, the presence of medullary bone, a special type of temporary bone tissue unique to reproductively active female birds, was identified in one of the presumed female specimens. By contrast, none of the males showed any evidence of medullary bone. Not only does this support the theory that *Confuciusornis*

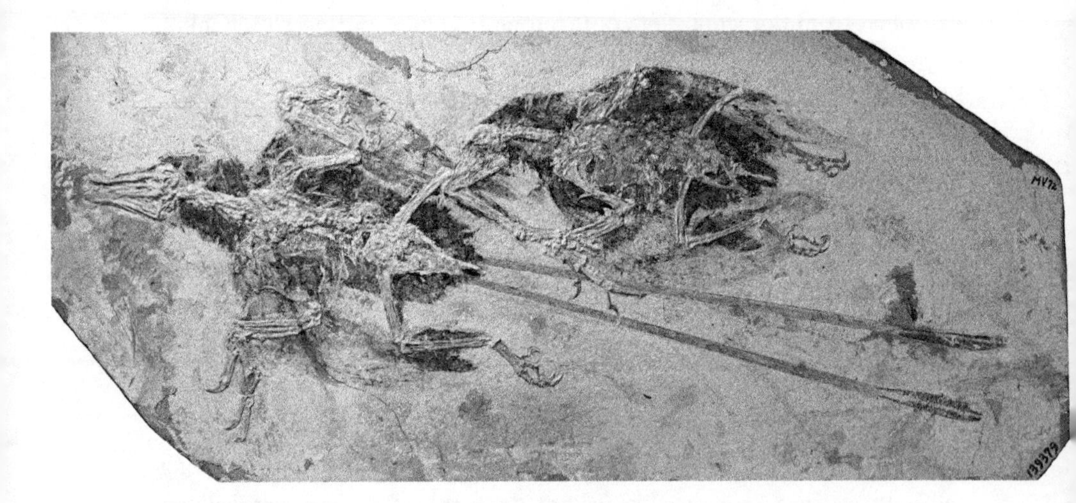

FIGURE 1.14. Two exceptional specimens of *Confuciusornis sanctus* with beautifully preserved feathers, with the male (*left*) displaying the extremely long streamer-like tail feathers and the female (*right*) without.

(Courtesy of Yongdong Wang, Nanjing Institute of Geology and Palaeontology)

BOB NICHOLLS 2020

individuals without tail feathers are female and those with them are male, but it suggests that the female individual with preserved medullary bone was either ovulating and getting ready to lay her eggs before she died or that she had just laid her eggs and died shortly thereafter. Medullary bone was not found in the other females because they were not in reproductive mode when they died.

Additional features that might highlight sexual differences in *Confuciusornis* revolve around the color and pattern of their feathers. That might sound like it comes from the realms of fantasy, but studies of some *Confuciusornis* feathers have helped to show that it was dark, possibly gray or black, all over the body and with a lighter coloring on the wings. One astounding specimen even revealed a suite of complex patterns, including small spots on the wings, crest, and throat. Although further study is required, it is tantalizing to ponder whether the color and patterns are indicators of sexual display or camouflage, as they typically are in birds today.

The spectacular preservation and abundance of *Confuciusornis* fossils provides us with a rare and fascinating look at sexual dimorphism in prehistoric birds 125 million years ago. Without the presence of feathers, the distinction between males and females might never have been demonstrated. Compared with living birds, the difference in plumage strongly suggests that the highly elaborate tail feathers of male *Confuciusornis* served an important purpose in sexual display.

FIGURE 1.15. (Overleaf). *His best moves.*

A male *Confuciusornis sanctus* struts his stuff as the female checks out his moves.

SHELL-SHOCKED MATING TURTLES

One of the greatest thrills of fossil hunting is knowing that there is always a chance that you could discover something entirely new to science. You can never quite predict what you might find, but even the most seasoned fossil hunters would have been shocked by the discovery of a pair of 47-million-year-old mating turtles.

This is the first and oldest unequivocal act of vertebrate sex in the fossil record—that is, two animals found actually having sex. It really is quite a find, considering how slim are the chances of both partners dying while mating, staying intact, and then being preserved as fossils. The conditions had to be perfect.

These turtles lived in a prehistoric lake that was once surrounded by a lush tropical forest. Known as Messel Pit, located near Frankfurt in west-central Germany and formerly the site of an oil shale mine, it is now a UNESCO World Heritage Site—one of more than a thousand unique and globally recognized sites designated by the United Nations' Educational, Scientific and Cultural Organization. Although small in diameter, Messel Pit was a deep, roughly three-hundred-meter volcanic maar lake that became a death trap, preserving thousands of plants and animals. The upper layers of the lake were inhabitable, where animals swam freely, but if they ventured—or fell—into the toxic bottom waters built up by volcanic gasses and decaying matter, they perished.

Among the countless fossils discovered at Messel, more than ten pairs of female and male turtles have been found, and all belong to the dinner-plate-sized species *Allaeochelys crassesculpta*. These pairs are either directly in contact or separated by no more than 30 centimeters, and all have their rear ends pointing toward each other.

Turtles are very ancient animals whose origins stem back at least 240 million years. The earliest species had teeth, rather than a toothless beak such as in modern turtles, and some even lacked the very thing that we have come to associate with them: a shell. By looking at living species, it is possible to decipher the sex of the *Allaeochelys* pairs. As in most living turtles, including the aptly named pig-nosed turtle (the closest living relative of *Allaeochelys*), which is found primarily in Papua New Guinea and northern

Australia, males have longer tails that extend beyond the shell rim, whereas females have shorter tails that barely reach the rim. The same difference in tail length can be seen among the Messel pairs. These distinguishing characteristics indicate that all of the fossilized males are on average 17 percent smaller than the females, which also display evidence of a movable shell hinge, which might have helped when laying eggs.

In two of the pairs, the tail of the male is aligned and wrapped below the female's shell, where they are in direct contact. This is the exact position where the tails are held when living turtles mate. All living aquatic turtles mount and mate in the water, where they often freeze in this position and sink through the water column before separating. Bearing this in mind, *Allaeochelys* belongs to a group of turtles that have skin porous enough to

FIGURE 1.16. A pair of *Allaeochelys crassesculpta* turtles in the middle of mating, with the male on the left and female on the right.

(Courtesy of SGN. Photo by Anika Vogel)

BOB NICHOLLS 2020

enable them to absorb oxygen from the water. Thus, like modern species, the male *Allaeochelys* probably mounted the female at the water's surface and began mating. As the pair froze in this embrace, we can theorize that they accidentally sank deeper into the abyss, where their porous skin absorbed the poisonous toxins of the lake, resulting in their deaths.

Love and death by poison; there is something a little Romeo and Juliet–like about this story. Representing the first and only known pairs of mating vertebrate fossils in the world, these specimens capture a truly intimate moment shared by prehistoric turtle couples.

FIGURE 1.17. (Overleaf). *Their last embrace.*

Getting it on in the Eocene Messel Lake, two *Allaeochelys crassesculpta* turtles engage in sex.

TINY HORSES AND TINY FOALS

It is impossible to pick up a book on fossils and evolution without reading about the evolution of the horse, which went from a house cat–sized, multi-toed browser to the modern-day, single-hoofed grazer. Identified from fossils found in Wyoming, the earliest tiny horses appeared almost 56 million years ago, although the most complete and exceptional early horse fossils come from two slightly younger (44–48-million-year-old) sites in Germany, at Messel Pit and Eckfeld.

We already know how extraordinary Messel fossils are from the pairs of mating turtles that have been found, and Eckfeld is similar. The sites are perhaps most famous for the preservation of spectacularly complete mammal fossils, often found with soft tissues, including fine details of skin and hair and sometimes even internal organs. One of the mammal species known from such fossils is a fox-sized horse called *Eurohippus messelensis*, which had four toes on its front feet and three toes on its back feet.

At Messel, there are four species of fossil horse, and *Eurohippus* is by far the most common, known from more than forty skeletons, the most of any early fossil horse found anywhere in the world. It is the smallest of Messel horses, at no more than 35 centimeters at the shoulder. Several of the specimens have been found with their body outline present, surrounding the skeleton as a sort of silhouette; some even preserve the outer ears and show that their rather short tails had a puffy tassel near the end. If that is not amazing enough, in 1987, a pregnant *Eurohippus* with a tiny well-preserved fetus inside was described. Remarkably, eight pregnant *Eurohippus* mares have now been identified from Messel, and a pregnant mare of a slightly larger species, *Propalaeotherium voigti*, was found at Eckfeld.

Just like living horses, which typically give birth to a single foal, each of the pregnant fossil mares contain just one fetus. The fetuses differ in their completeness, some with all of the bones intact, whereas others are jumbled up, and several have well-developed milk teeth. All of the mares are in the later stages of pregnancy; a foal would have been around 15–20 centimeters long at birth—easily able to fit in your lap. Interestingly, some of the pregnant mares retain their milk teeth as well, meaning that these individuals were not fully adult when they became pregnant.

Eurohippus mares also had a wide pelvic channel, like modern mares, whereas the pelvic channel in stallions was narrow, so, without the presence of a fetus, this feature can be used to distinguish the sexes. Inevitably, it is difficult to estimate the gestation period for *Eurohippus*. In modern horses, the norm is eleven months, but the length of gestation in living mammals is influenced by body size, with larger-bodied mammals having longer gestation periods. Taking the small size of *Eurohippus* into account, it is assumed that the gestation period might have lasted at least two hundred days (or six and a half months). This is all very impressive, but the most stunning discovery with these pregnant mares has to do with the soft parts.

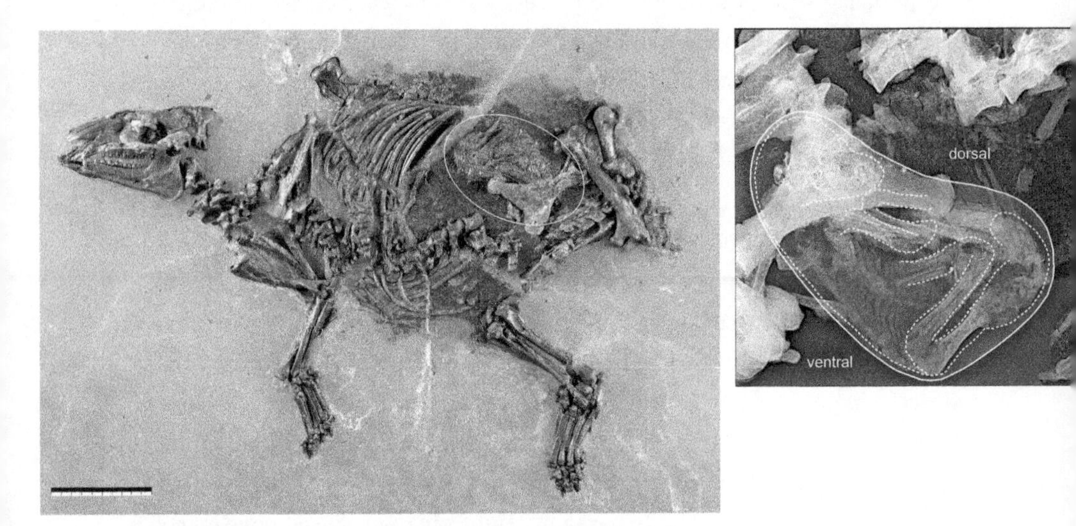

FIGURE 1.18. Pregnant *Eurohippus messelensis* with a single unborn foal preserved inside (*left*); X-ray close-up of tiny, almost complete foal and reconstruction of the original position of the fetus (*right*).

(Images from Franzen, J. L., et al. 2015. "Description of a Well Preserved Fetus of the European Eocene Equoid *Eurohippus messelensis*." *PLOS One* 10, e0137985.)

FIGURE 1.19. (Opposite). *Love in the undergrowth.*

A herd of *Eurohippus messelensis* walks through the forest undergrowth at the edge of the Messel Lake as a dominant stallion mounts a mare.

BOB NICHOLLS 2020

The Eckfeld *Propalaeotherium* was the first to reveal such amazing details. Mind-blowingly, the fetus in this pregnant mare was still covered by parts of a placenta, the first report in a fossil mammal. But of all the pregnant mares, by far the most astonishing was a specimen excavated at Messel in 2000 by a team from the Senckenberg Research Institute. Inside the mare's pelvis area is a tiny exceptional fetal skeleton that is almost entirely articulated and complete except for the skull, which is crushed, but the soft tissue preservation is unrivaled.

To interpret the preserved soft parts, the specimen was examined with a powerful scanning electron microscope (SEM) and was subjected to a high-resolution micro-x-ray, which revealed the unprecedented find of a reproductive tract. The fetus was still enwrapped in the uteroplacenta (the uterus and placenta) and is attached to the broad ligament, a structure that connects the uterus to the lumbar vertebrae and pelvis and helps to support the developing fetus. The extent of the uteroplacenta, as well as the presence of wrinkling across its surface, is identical in living mares, and this fossil represents the earliest record of the uterus of a placental mammal. The fetus was nearly full term when the mare died, but its position, being upside down rather than right side up, as in living horses when they go into labor, suggests that the mare did not die during birth but for some other unknown reason.

The occurrence of multiple *Eurohippus* fossils at Messel suggests that they might have lived in small groups or perhaps even herds, providing possible information about parental care. In spite of the obvious differences in size and skeletal features, these exceptional and tiny pregnant horses show how the reproductive anatomy and corresponding behaviors have changed very little over the last 48 million years.

2 Parental Care and Communities

A dead, decaying mouse emits a pungent odor on the forest floor, filling the air with chemicals and attracting the attention of burying beetles, inch-long undertakers of the animal world, quick to arrive at the "fresh" carcass and claim it for themselves. Tag-teaming males and females pair up and fend off rivals before moving and burying the mouse in their underground chamber, where they strip away the fur and preserve it with special secretions. The female lays her eggs near the carcass, and, in a rare example of insect parental care, the male sticks around to help the female, guarding the crypt and the brood. Using the mouse as a nursery to rear their baby burying beetles, both parents also use it to feed their larvae. Cleaning the forest floor of its dead carrion, these beetles are rotting carcass connoisseurs, recycling a variety of small dead vertebrates, including other mammals, birds, and even snakes. In death, these small animals breed new life.

This is probably not the example of parental care you were expecting. Some might say it is pretty gross (not me). It is not quite what springs to mind when we think of animals caring for their young, such as a protective mother hen nurturing her cute fluffy chicks. Instead, this example shows the bizarre and often highly complex behaviors that even some insects have developed in order to give their offspring the best chance of survival.

The extent of parental care varies tremendously across the animal kingdom. New examples are recorded every year, just like the recent discovery that the world's largest frog (the Goliath frog) builds nests out of rocks to protect its developing young. Although some species express no parental care at all, like many invertebrates that lay eggs and move on, other, more complex animals such as primates invest years in raising their young.

Parents provide care in a variety of ways by commonly building nests, feeding, and protecting their offspring from would-be predators. Some species, like the burying beetles, have developed unusual and often extreme ways to care for their young. Take the group of fish known as mouthbrooders. These fish cease feeding for extended periods to protect their young by brooding them inside their mouths. Once the eggs are hatched, the tiny free-swimming fry seek shelter inside the parent's mouth for up to several weeks. Parents often sacrifice a lot for their offspring, and lengthy, restrictive periods of parental care can sometimes have detrimental or even deadly effects on the parent(s).

Many animals live in communities. Some species stay together in groups year round, such as immense penguin colonies or close-knit elephant herds, whereas others form only temporarily and meet at particular times (like during mating season). Some species of burying beetles even occasionally share carcasses, burying them together and rearing their offspring communally, after which they separate. There can be many benefits of living in communities, such as added protection and an increased chance of finding mates and food. On the flip side, there is more competition and an increased risk of disease, among other things. Communities can provide a wealth of information about an ecosystem as a whole, how different species interact; compete, rear, and protect their young; form complicated social structures; and much more.

Having an understanding of parental care and community life in living animals gives us a starting point to help interpret what it might have been like for their prehistoric predecessors. For sure, just like animals today, many prehistoric species would have cared for their young, but figuring out how they did this can seem, well, impossible.

To emphasize this difficulty in the fossil record, imagine finding a fossil mouthbrooder with tiny fry in its mouth or burying beetles inside a carcass. Do you think we could reliably interpret what was happening? Maybe. But it would require careful assessment of the fossil and comparison with living analogues, along with multiple examples, if available.

Contrary to what we might think, prehistoric communities can be a little *easier* to understand. Sort of. Paleontologists can take advantage of fossil-rich sites, many of which are referred to as *Lagerstätten*. This is not a fancy German beer but a word that describes fossil sites of exceptional detail, those that are known for their great abundance of fossils, exquisite preservation of soft parts of organisms that ordinarily would not be preserved as fossils, or a combination of both; examples of which include the Messel Pit (see chapter 1) and the Burgess Shale (later in this chapter). These provide an excellent opportunity to understand the paleoecology because they often preserve a wide range of fossils that can help to paint a picture of how these animals might have interacted. Hundreds or even thousands of plant and animal species might be found in a particular formation, and these can be used to help to interpret what the environment was like. Based on the fossils, paleontologists can make an educated guess about which animals lived in herds, which were the top predators, and so forth. Even then, with such fossiliferous sites, they still (often) provide only clues and not quite direct evidence.

In order to reliably interpret such behaviors, it would seem that we need direct evidence of, say, parent(s) and their eggs or babies preserved together, or perhaps an abundance of animals found in exactly the same rocks interacting in some way. Well, what about when we *do* have these fossils? In this chapter, we take a deep dive into a select few extraordinary fossils that provide a little insight into ancient parental care and interacting animal communities showcasing behaviors that seem familiar yet are buried deep in time.

BROODING DINOSAURS

The Gobi Desert in Mongolia is a hotspot for dinosaur discoveries. In 1923, an expedition led by a team from the American Museum of Natural History to the famous Flaming Cliffs fossil site resulted in the discovery of several new species. Among them was a strange-looking theropod with a long neck, short skull, and toothless beak. It was described a year later by the museum's paleontologist and president, Henry Fairfield Osborn, who also took part in the expedition. Bizarrely, its incomplete skeleton was found lying over a nest of dinosaur eggs. Osborn assumed that the eggs belonged to a small ceratopsian dinosaur called *Protoceratops andrewsi* because skeletal remains and eggs of this species were commonly found in the same area. He named this new theropod *Oviraptor philoceratops*, meaning the "egg seizer with a fondness for ceratopsian eggs."

Apparently, this *Oviraptor* had been caught in the act of robbing a dinosaur nest, feasting on the eggs of another species. This would be a mega find, a world first, but it was incorrect. *Oviraptor* was no egg thief. At least, not this one. In a huge dose of irony, in 1994, seventy years after the description of the fossil, a reexamination of the original eggs, along with a comparison with new specimens, showed that the eggs belonged to none other than *Oviraptor*, not *Protoceratops*.

You are probably familiar with this story of *Oviraptor* the egg thief. For me, growing up, it was one of those mainstream so-called dinosaur facts that seemed to be everywhere— in books and documentaries and splashed across museum displays, and occasionally it pops up even today. Funnily enough, Osborn himself thought that the name (*Oviraptor philoceratops*) might end up painting a false picture of the animal's feeding habits. He was correct!

With the misidentification realized and *Oviraptor*'s reputation restored, evidence indicated instead that this was a parent who was sitting on its nest, perhaps brooding and guarding the clutch of eggs. Backing up this claim was a spectacular fossil find in 1993, during a joint expedition to the Gobi by the American Museum of Natural History and the Mongolian Academy of Sciences. This trip led to the discovery of a new, extremely fossil-rich locality called Ukhaa Tolgod. It was here that a skeleton of an emu-sized, *Oviraptor*-like dinosaur was also found sitting on its nest. This dinosaur, although closely related to *Oviraptor*, being a member of the same family, was later identified as an example of a new genus and species, called *Citipati osmolskae*.

Sitting in the center of the nest, with both hind limbs tightly folded and the feet and lower legs almost parallel to each other, the skeleton is preserved in the same position that it died. Its arms, which were almost certainly feathered, are wrapped around the nest, lying across several eggs. At least fifteen eggs are preserved, each measuring 16 centimeters long by 6.5 centimeters wide, and are arranged in pairs in a circular pattern. The arrangement of the eggs suggests that the parent(s) might have moved

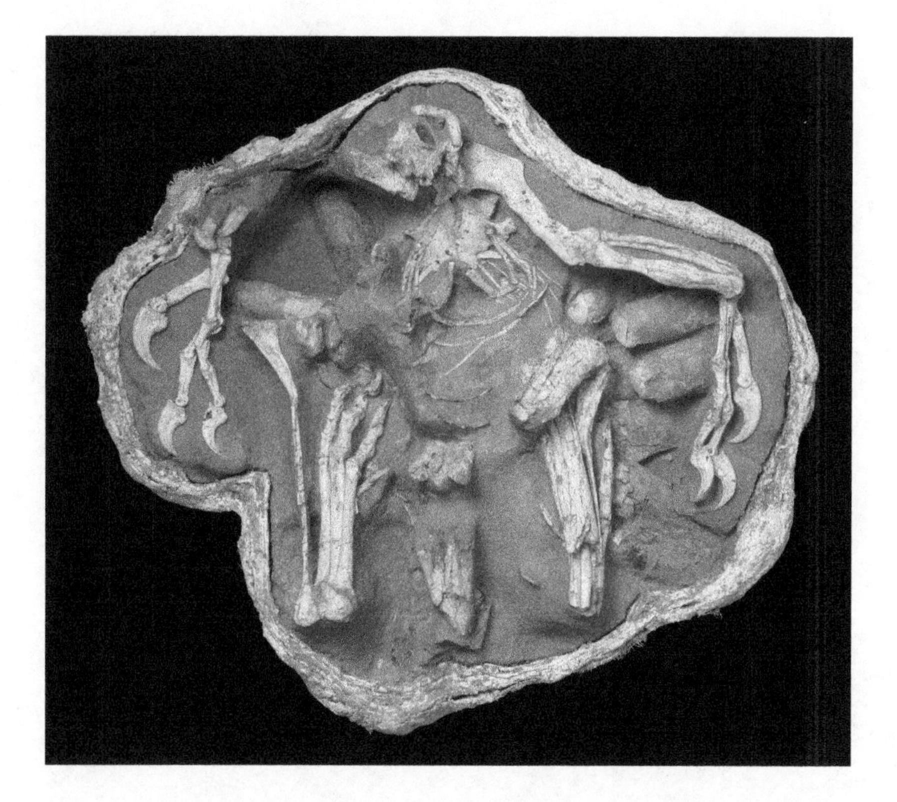

FIGURE 2.1. The oviraptorosaur, *Citipati osmolskae* ("Big Mama"), sitting on a nest of eggs.

(Courtesy of Mick Ellison, American Museum of Natural History).

BOB NICHOLLS 2020

them, or that the mother positioned herself to lay the eggs in a particular manner. It has been found that, like crocodiles, some oviraptorosaur dinosaurs had two functional oviducts, so it is plausible that two eggs were laid at the same time, placed side by side.

This *Citipati* was incubating its eggs, keeping them warm and protected, just as modern birds do when brooding. It has been nicknamed "Big Mama," although there is nothing to suggest a female identification, and it could well have been "Big Dada." One study even argued for the latter and indicated birdlike paternal care in these dinosaurs. Brooding behavior differs among living birds, and both the males and females usually take turns sitting on the eggs, although in some species, only one of the sexes cares for them. It is possible that oviraptorosaurs displayed similar behavior, and the other parent might have been nearby.

A startling eight oviraptorosaurs preserved with their nests have now been found, including a second specimen of *Citipati* on its nest. Five are from the Gobi and three are from southern China, and all are around 70–75 million years old. The most recently discovered specimen was found to preserve a clutch of at least 24 eggs, including seven that contain embryonic remains. Each dinosaur is preserved in the same position that it died, rapidly buried by violent sandstorms and immense sand slides, and they represent some of the rarest of fossils.

The changing interpretation of *Oviraptor*, from egg-stealing villain to caring parent, has helped to reveal an extraordinary story and highlights the importance of finding additional specimens to answer questions about prehistoric behavior. The birdlike parental care demonstrated by these dinosaurs supports a behavioral link with modern birds and shows that this brooding practice evolved long ago.

FIGURE 2.2. (Opposite). *A stand against the sand.*

A male *Citipati osmolskae* carefully sits on the nest as a female lingers in the background. An enormous sandstorm approaches in the distance.

OLDEST PARENTAL CARE: ANCIENT ARTHROPODS AND THEIR YOUNG

The beginning of the Cambrian period marked one of the major transitions in the history of life on Earth. The "Cambrian explosion," referring to the sudden (in geological terms, meaning that it occurred over several million years) appearance of more complex animals with mineralized skeletons, has been covered extensively in books, popular writings, and numerous TV documentaries. And with good cause. The rich influx of animals—representing the first appearance of many animal phyla and basic body plans known today, including other forms that even paleontologists are uncertain of—resulted in ecosystems' becoming more complex. The oceans were teeming with life; the first animals with legs and even compound eyes appeared, each differing greatly in size, shape, ecology, and lifestyle and ultimately evolving novel ways of reproducing. This was a time of great change and experimentation.

It "appeared as if it were still alive on the wet surface of the mudstone," wrote Hou Xian-guang in his field notebook on July 1, 1984, while examining a fossil he and his team had found in Yunnan Province, southwest China. Although fossils had been reported in the area before, the purpose of this trip was to illuminate their significance. Known as the Chengjiang Fossil Site, this location is so exceptional that it preserves not only the hard parts of animals but also spectacular details of their entire soft, squirmy bodies. These animals are around 520 million years old, from near the beginning of the Cambrian, and provide a remarkable snapshot of life during that time.

Among the first fossils studied by Xian-guang were the remains of arthropods, which at Chengjiang are particularly common in number and diverse in form. By far the most common is the seed-shrimp-like *Kunmingella douvillei*, which belongs to a group of extinct arthropods most closely related to crustaceans. This tiny, typically half-centimeter-long arthropod is identified by its two-part "butterfly" shell, a protective structure that covered most of its body and its ten pairs of appendages. Of the many thousands of known fossils, six were found with associated eggs, helping to shed light on the reproductive strategy of this mini arthropod. Members of this species clearly reproduced sexually, but they also cared for their eggs, increasing their chances of survival.

Between fifty and eighty eggs ranging in size from 150 to 180 micrometers were found attached to the legs of a female *Kunmingella*. To put their minute size into context, about six of these eggs could fit comfortably onto the tip of a ballpoint pen. That the eggs are attached to the legs in multiple specimens shows that this association is not a coincidence. Instead, this species had evolved a peculiar technique for storing its eggs, and it is possible that this behavior might have been driven by predation pressure. In particular, *Kunmingella* is one of the smallest Cambrian arthropods, and evidence from coprolites (fossil poo) shows that it was on the menu for larger animals.

Consequently, for such a species to survive in a radical new world filled with predators, it had to find a way of providing its offspring with the best possible start in life. In most species of living crustaceans, fertilized eggs stay attached to the females until they hatch. Given the great abundance of *Kunmingella* fossils, it is clear that this species was doing something right, and it seems likely that was due to how the eggs were stored and cared for.

These specimens are preserved in fine-grained mudstone and are thought to have been buried rapidly, smothered by mud. This suggests that they were found exactly where they died, meaning that the fossils had not been transported or disturbed, allowing for their fine preservation. The same mode of preservation is interpreted for another remarkable arthropod from Chengjiang that revealed a fascinating insight into complex parental care.

This shrimp-like arthropod, called *Fuxianhuia protensa*, is known from specimens of varying size spanning a range of growth stages, from very young to old adults, which provides an overview of the differences in *ontogenetic development* (how the animals change as they age). Although examples from each ontogenetic stage share a similar body structure, notably, the smaller individuals have fewer body segments (called *tergites*) than the older individuals. This shows that additional segments were added to the body (through molting) as the animal aged, a process called *anamorphosis*, which is found in living arthropods such as millipedes. So, it is possible to deduce the age of a specimen, in addition to their size and shape, by counting the individual segments.

In one exceptional *Fuxianhuia* specimen, a large (8 centimeters) sexually mature adult was found alongside four very small (around 1 centimeter) juvenile individuals of the same size and age, determined by the number of body segments. Given their association and the fact that the specimens were buried together at the same time, evidence suggests that the four individuals are from the same clutch and that the adult is the parent, caring for its developing offspring. This strategy of extended parental care can enhance the chances that the offspring will survive and thus continue the family line.

Kunmingella and *Fuxianhuia* are not the only Cambrian arthropods to provide such evidence. In September 1909, while exploring the breathtaking Canadian Rockies, Charles Doolittle Walcott discovered the Burgess Shale. This legendary fossil site, famously popularized by Stephen Jay Gould in his 1989 book *Wonderful Life*, was the first to yield exceptional Cambrian fossils, the detail and diversity of which had never been seen before. Much of what we know about Cambrian life comes from studies of the Burgess Shale material, which is slightly younger than Chengjiang, around 508 million years old. Walcott amassed an enormous collection of around 65,000 specimens and named several new species, including *Waptia fieldensis* in 1912, named after Mount Wapta and Mount Field, two mountains near where the fossils were collected. Having a striking resemblance to modern shrimp, *Waptia* possesses a long tail, has a maximum size of about 8 centimeters, and is known from several thousand fossils. Despite having first been described over a century ago, only recently, in 2015, were the reproductive habits of this arthropod revealed.

After an extensive study of 1,845 *Waptia* specimens, 5 were seen to have eggs hidden inside. Positioned on either side of the body, small clusters of up to twelve eggs were found tucked under the carapace near the head, suggesting that as many as twenty-four eggs could be preserved with a single female. Each egg was relatively large, with an average diameter of

2 millimeters, but, most remarkably, they contained tiny, developing embryos. This represents the oldest direct evidence of a mother associated with her eggs and embryos.

Waptia and *Kunmingella* differ substantially not only in size and structure but also in clutch number and egg size, as well as how the eggs were

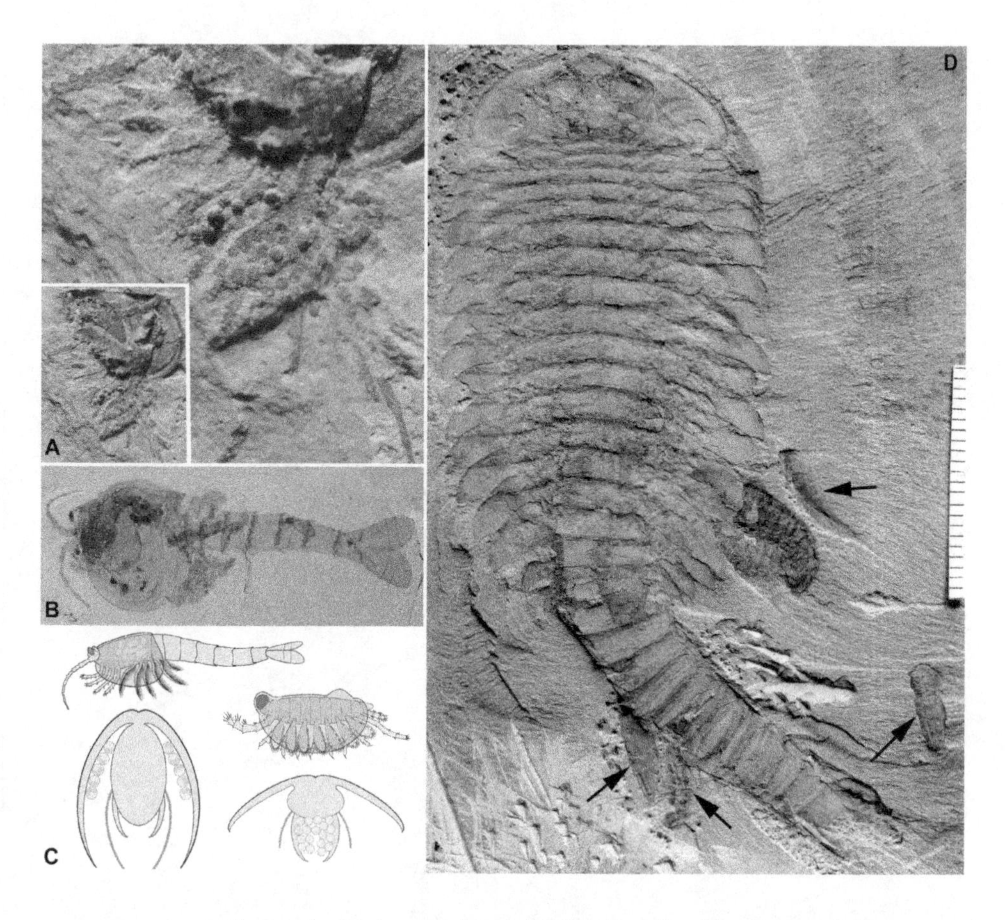

FIGURE 2.3. Early arthropod parental care: (*A*) *Kunmingella* with eggs attached to its legs; close-up to the right. (*B*) *Waptia* with clusters of eggs under the carapace. (*C*) Reconstruction showing where eggs were kept on the former two species. (*D*) *Fuxianhuia* adult with associated tiny juveniles (identified by the arrows).

([*A*] Courtesy of Jian Han; [*B–C*] courtesy of Jean-Bernard Caron, Royal Ontario Museum, illustration by Danielle Dufault, Royal Ontario Museum; [*D*] courtesy of Dongjing Fu)

BOB NICHOLLS 2020

attached to and cared for by the mother. The *Fuxianhuia* fossil provides the oldest evidence of extended parental care in the fossil record. These three different modes of parental care highlight how complex behavioral strategies had already evolved among species more than half a billion years ago.

FIGURE 2.4. (Opposite). *Eonurture.*

An adult *Fuxianhuia protensa* cares for several babies, which stay close by her side in a sheltered area on the seafloor.

PTEROSAUR NESTING GROUNDS

Pterosaurs are among the most unusual and awe inspiring of prehistoric creatures. They were the first vertebrates to evolve powered flight, a feat only achieved later by birds and bats. Before the appearance of pterosaurs during the Middle Triassic, roughly 220 million years ago, insects had long dominated the skies. Despite more than two centuries of detailed study, the early lives of these winged reptiles remained puzzling, but that recently changed with the discovery of one of the most significant pterosaur finds ever made. No "*egg*xaggeration."

It had always been assumed that pterosaurs, being reptiles, must have laid eggs, presumably in nests, but direct evidence had remained elusive. Then, like buses—or, perhaps a better analogy today would be phone apps—you wait for one to come along, and three appear together.

In 2004, three beautifully preserved eggs containing tiny pterosaur embryos were described, confirming that pterosaurs were indeed egg layers. Two of the eggs were collected from Cretaceous-aged rocks in the Jingangshan area of Liaoning, China. One of them was so well preserved that it was found to have a soft, leathery shell, just as with most modern reptiles. A few years later, another surprise came from an unlikely source, a local farmer, who had collected another egg, but this time from Jurassic rocks in Liaoning. Eradicating any doubts concerning the egg-laying ability of pterosaurs, this one was still connected to the mother.

Nicknamed "Mrs T," this is a virtually complete specimen of the pterosaur *Darwinopterus*, named after Charles Darwin. The fossil is preserved as a part and counterpart of the skeleton, with one side being more complete than the other. Mrs T's oval-shaped egg is positioned between her legs, where it is associated with the pelvis near the base of the tail. (The egg measures just under 3 centimeters long and is estimated to have weighed 6 grams.) This position suggests the egg was squeezed out of the body during decomposition, due to a build up of gases from within. Through subsequent study of the less complete slab, a second egg was found inside the mother, revealing that pterosaurs, like most living reptiles, had two functional oviducts. It is plausible that Mrs T was readying to lay her eggs shortly before she died.

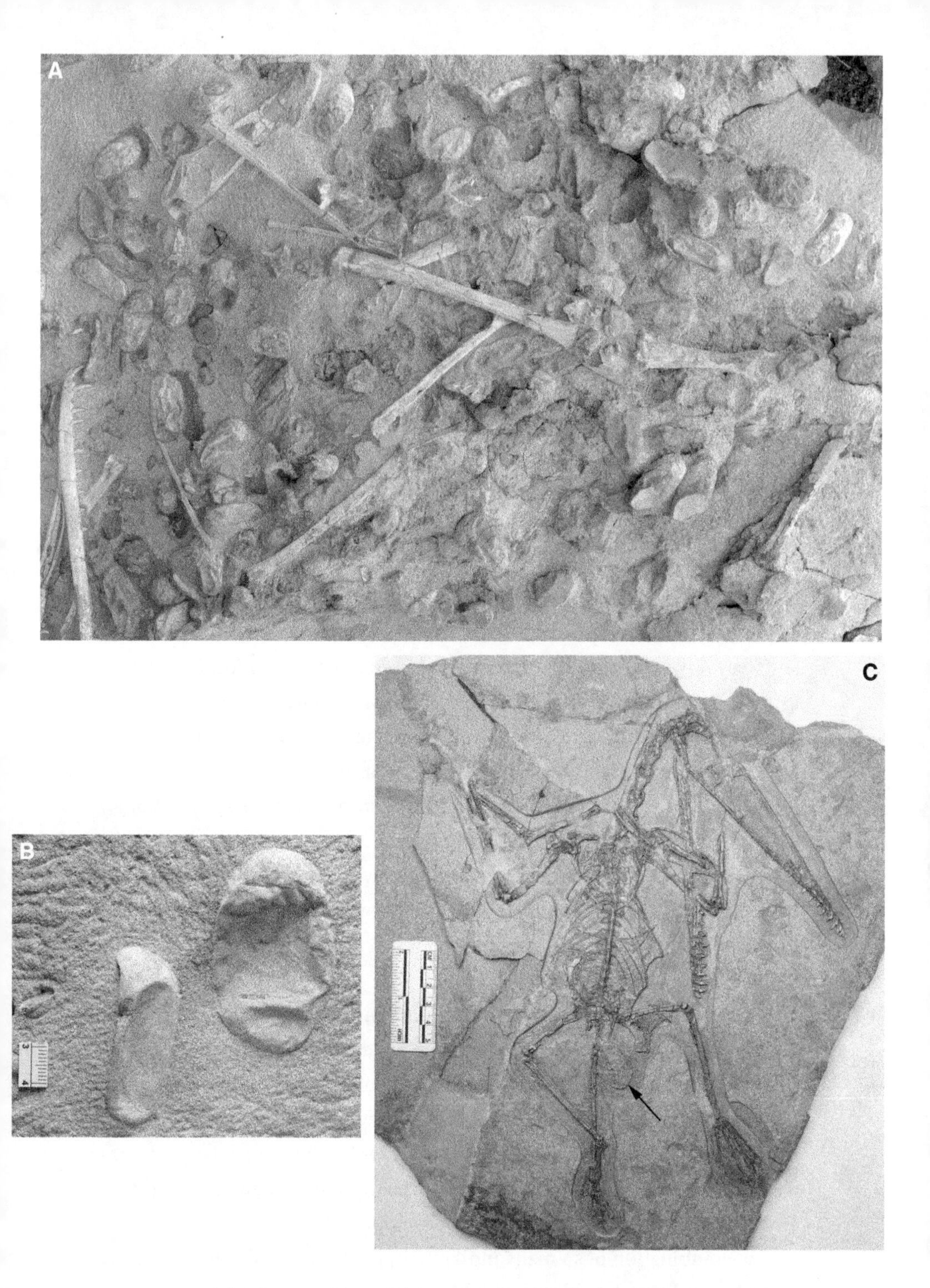

FIGURE 2.5. (*A*) A large portion of the mass assemblage of *Hamipterus tianshanensis* bones and eggs from the Turpan-Hami Basin site. (*B*) Close-up of two exceptional eggs. (*C*) Female *Darwinopterus* ("Mrs. T") with arrow pointing to the complete, attached egg.

([*A–B*] Courtesy of Xiaolin Wang and Wei Gao; [*C*] courtesy of Dave Unwin)

This pterosaur fossil was also the first to confirm sexual differences between male and female pterosaurs. Namely, this mother lacked a head crest and possessed a large pelvis, the opposite of that found in other *Darwinopterus* specimens. The latter had been presumed to represent males of the species, but only now, with the egg–mother association, can this assumption be confirmed and these differences identified as sexually dimorphic.

Such Chinese discoveries are really exciting and have helped to fill many gaps in our understanding of pterosaur behaviors. But none has provided as much information as a 120-million-year-old discovery made near the famed Tian Shan Mountains in Xinjiang, northwestern China. Since 2006, extensive fieldwork has been carried out in the Turpan-Hami Basin, just south of the mountains, and it is here that a new fossil site containing thousands of pterosaur bones representing both males and females and hundreds of eggs, several with three-dimensional embryos, have been found together in a mass mortality assemblage. Similar to *Darwinopterus*, the males and females differ in the size, shape, and robustness of the head crest, males having the far larger, more prominent crests of the two.

At least forty individuals were identified from the bone bed, although it has been suggested that the actual number might be in the hundreds. These skeletons were found to represent a new species, named *Hamipterus tianshanensis*, which had a maximum wingspan of 3.5 meters, comparable to that of the wandering albatross, which has the longest wingspan of any living bird.

A staggering three hundred-plus *Hamipterus* eggs were found at this site. Although slightly deformed, having been squashed, they are three-dimensionally preserved and display exceptional detail. The eggs also bear numerous small cracks, suggesting that, although they are soft and leathery, a very thin and delicate mineralized shell was present. A total of forty-two eggs were found with contents intact, and sixteen of them contained identifiable embryonic bones in varying stages of development.

The combination of so many individuals, including adults of both sexes, juveniles and tiny hatchlings, along with the huge number of eggs, indicates that a nesting site was nearby. It further suggests that this species was gregarious, living in large groups and nesting in colonies, perhaps on the shores of lakes or rivers. Similar to many reptiles, it seems that *Hamipterus* buried its eggs in the sand, to prevent their drying out. That multiple specimens of both males and females have been found might suggest that they stayed near the nests and helped to rear their young, protecting them from predators.

It appears that the nesting site was struck by powerful, high-energy storms that dislodged the eggs from their nests and washed them into a nearby lake, where they were buried alongside the pterosaur skeletons. In light of the general rarity of pterosaur fossils, lack of associated skeletons, and rarely found eggs, this mass mortality is a truly spectacular discovery that provides an unprecedented insight into the lives of these winged wonders.

FIGURE 2.6. (Overleaf). *The windswept colony.*

Chaos ensues at the pterosaur nesting ground as a harsh storm disturbs the *Hamipterus tianshanensis* colony and their eggs, causing some of the eggs to roll down the sandy shore and into a lake.

BOB NICHOLLS 2020

MEGALODON NURSERY

Ask any fossil fan to name you an extinct apex predator, and "Megalodon" will be near the top of the list. That is no surprise considering that this megatooth giant is the largest shark species known to have existed, at around 16 meters in length, more than twice the size of the great white shark in the movie *Jaws*. Yes, this shark was *that* big. Its gigantic size lends it mass public appeal. People are always interested in knowing which was the largest or ultimate predator, and in recent years, there has been considerable interest in Megalodon in both the scientific and popular realms.

With all the hype surrounding its size, it is understandable that other aspects of Megalodon's life can easily be, and are often, overlooked, considering that it had teeth as large as your hand and evidence that it dined on whales. Practically everything we know about Megalodon comes from their teeth, which are found in abundance around the globe, so it might be assumed that little information regarding their behavior, other than the obvious feeding, could be worked out. Be that as it may, a fascinating discovery made in Panama, Central America, has helped to reveal how Megalodon protected its young.

As a scientist, I would be remiss not to mention that the name "Megalodon" is taken from its scientific species name, *Otodus megalodon* (sometimes also called *Carcharocles megalodon*). So, really, we should refer to it as *Otodus* or *O. megalodon*, although that does not have quite the same appeal, right? Furthermore, with increased public interest in Megalodon, there has been some major speculation about it in the popular media—namely, that some believe it still exists today. It does not, although, in geologic terms, the species went extinct pretty recently, around 3.6 million years ago, so early hominids (human ancestors) might well have seen a living Megalodon, or at least one that washed up on the beach.

Located in the Isthmus of Panama, the narrow strip of land that connects North and South America, is a marine fossil formation known as the Gatun, which has yielded a rich and diverse fossil shark fauna. Ten million years ago, during the Miocene, this area was a warm shallow-water ecosystem roughly 25 meters deep connecting the Pacific Ocean and the Caribbean Sea, which today are separated by the isthmus.

Although shark fossils are frequently found in the Gatun, Megalodon teeth are not very common. During fieldwork between 2007 and 2009, twenty-eight Megalodon teeth were collected from two fossil locations in northern Panama, at Las Lomas and Isla Payardi. The teeth were studied by a team from the Florida Museum of Natural History led by paleontologist and Megalodon expert Catalina Pimiento. An additional twenty-two teeth were collected during another field trip, but only twelve of them were complete enough to be studied. Surprisingly, most of the teeth were small to very small, some having a crown height of less than 2 centimeters, and large examples were uncommon.

In Megalodon's jaw, the teeth vary in size, and therefore the small size might be argued to represent their relative position in the jaw. Testing this theory, the team compared the teeth with known sets of associated Megalodon teeth of different life stages (juvenile and adult) from localities in Florida and North Carolina. Most of the Gatun teeth were found to be from juveniles and neonates (newborns) with estimated total body lengths between 2 and 10.5 meters; the smallest individuals were considered to represent embryos. Although some of these older juveniles were very large (6–10 meters), they would still have been on the menu for adult Megalodon. The highly unusual presence of so many small teeth belonging to embryos, newborns, and juveniles supports evidence that the Gatun was an ancient Megalodon nursery.

In many modern sharks, juveniles and newborns are commonly found in nursery areas, often living alongside several shark species. These nurseries provide an essential habitat for the young sharks, which are particularly vulnerable to predators (mainly larger, adult sharks), and can offer protection and ample food resources, increasing the youngs' chances of survival.

In the Gatun paleo-nursery, various types of bony fish and other sharks are known and represent a food source for the young Megalodon. With the threat of gigantic adults, as well as other large sharks, the juveniles presumably hunted large marine mammals only when they reached adult size. (Incidentally, marine mammal fossils are uncommon in the Gatun.) Similarly, juvenile great white sharks mostly feed on fishes (including other sharks) and transition to feeding on mammals during their adult stage. Although not a direct relative, the great white can act as a good ecological

1cm

analogue for Megalodon due to their similarities; namely, in their teeth and vertebrae and presumed body shape and feeding habits. Great whites also use nursery grounds.

Several large Megalodon teeth were found in the nursery, alongside the juvenile teeth, and represent adults with total body lengths greater than 10.5 meters. The presence of adults with the juveniles is not unexpected. Sharks constantly replace their teeth throughout their lifetime, so it is plausible that when a female Megalodon laid her eggs or gave birth to live young in the nursery area, some of the teeth easily could have been dislodged and lost. In addition, although nurseries can be places of refuge, there is no guarantee that large individuals will stay out of the area, and therefore some of the juveniles might have been preyed on by large adults.

Ten years after Pimiento's team announced their discovery, in 2020 a different group of researchers reported a potential Megalodon nursery from a newly recognized site in the province of Tarragona, Northeastern Spain. The team also reexamined several other well-known fossil sites and formations containing Megalodon teeth, and, based on the high proportion of individuals found within the range of neonates and young juveniles, established three more that were consistent with nursery grounds (in addition to Gatun and Tarragona). Two are located in the United States—in Maryland and Florida—and the other is at a location in Panama. Together, all five sites range in age from 4.7 to 15.5 million years.

These findings reveal that nursery areas were commonly used by Megalodon over a wide geographic area for millions of years, playing a pivotal role in the survival of the species. The identification of Megalodon nurseries provides a very different view of the behavior of this most impressive super-predator.

FIGURE 2.7. (Opposite). A selection of *Otodus megalodon* teeth collected from the Gatun Formation, ranging from juveniles to large adults.

(Image modified slightly from Pimiento, C., et al. 2010. "Ancient Nursery Area for the Extinct Giant Shark Megalodon from the Miocene of Panama." *PLOS One* 5, e10552.)

FIGURE 2.8. (Overleaf). *Beyond the reach of mega-teeth.*

Numerous juvenile *Otodus megalodon* seek safety in the shallows of the nursery ground as a large adult swims by in the deep water.

BOB NICHOLLS 2020

THE BABYSITTER

If you have a child, at some stage or another you will need a babysitter. When that time comes, you might call on your parents, your siblings, or perhaps that friend who owes you a favor. Humans are not the only ones to rely on babysitters; hundreds of animals do, too. Whether fairy wrens, meerkats, lions, or whales, members of each group help to raise offspring that are not their own.

Having somebody else temporarily take care of your child might be daunting, though it gives you the necessary time to procure food, socialize, and rest, important factors in being able to readily provide for your offspring. In reality, babysitting is a common form of selfless cooperative behavior called *altruism*, whereby an individual will choose to help others at its own expense and often without any obvious reward (although money is usually the driving force for teenagers).

Finding reliable evidence for altruistic behaviors in fossils is problematic. Associations of large and small individuals of the same species might simply reflect a chance accumulation or perhaps another type of behavior. Therefore, several evidentiary factors must be present and the fossil preserved in such excellent condition that it can be interpreted with some degree of confidence. One such astounding fossil bone bed was collected from roughly 125-million-year-old Cretaceous rocks in China's Liaoning Province. It contained multiple, exceptionally preserved dinosaurs that were buried alive by an ancient volcanic debris flow (or *lahar*).

The dinosaur in question is *Psittacosaurus*, a Labrador-sized primitive bipedal ceratopsian. With hundreds to thousands of fossils recorded, it is by far one of the most common dinosaurs found across Asia. In one extraordinary case, a specimen was found with feather-like bristles on the tail and intact skin bearing distinct patterns. A detailed study of this fossil enabled scientists (including Bob Nicholls, the artist) to create the most lifelike and accurate reconstruction of a dinosaur ever, showing that it was dark on top and light underneath. This type of coloring pattern is referred to as *countershading* and is a common form of camouflage in modern animals, designed to hide the individual's three-dimensional shape. This suggests that *Psittacosaurus* lived in an environment with plenty of cover, such as a forest, which matches the types of plant fossils found in the same rocks that yield the skeletons.

The *Psittacosaurus* bone bed was formally described and officially unveiled in 2004, causing quite a buzz in the paleontology community. Measuring just 60 centimeters across, this small block revealed thirty-four articulated and similarly sized juveniles (with skull lengths ranging from 3 to 5 centimeters) tightly clustered with a single, much larger and clearly older individual of the same type. Each animal is preserved in three dimensions, laid in a lifelike upright pose with their heads slightly raised. Given their identical nature, the association led researchers to surmise that this was the parent and its offspring, thus evidence for post-hatching parental care. A major find.

Some paleontologists were not convinced by the fossil and questioned both its authenticity and the presumed behavior. One research team could identify only thirty juveniles, six of which were represented by their skull alone. Although of similar size and preservation as the rest, these skulls were not completely encased in rock. Sadly, there is a dark side to the private fossil trade, whereby specimens are illegally collected, sold, and often modified. Allegedly, the collectors had added the skulls to the original block to enhance its "wow factor" and fetch a higher price. Whether or not this is the case, the team determined that only twenty-four juveniles definitely belonged together and were therefore genuine. The larger individual was also questioned, but parts of its skull and skeleton reside firmly embedded in the matrix and are intertwined with some of the juveniles.

In terms of the behavioral aspect, comparison of the large *Psittacosaurus* with other large skeletons from the same area showed that this individual was probably not yet an adult. For example, the skull length measures just 11.6 centimeters, almost half the size of the largest known *Psittacosaurus* skulls (which are up to 20 centimeters), whereas the smallest skulls measure just under 3 centimeters. Furthermore, in light of the abundance of *Psittacosaurus* fossils, a substantial body of work has been done on age determination using bone histology. Comparing these data with the studied specimen, this animal was approximately four to five years old when it died. The same studies have found that *Psittacosaurus* did not reach sexual maturity until at least eight or nine years of age. Therefore, the larger individual was too young to breed, meaning that the fossil cannot be an example of parental care and is instead that of an adolescent, perhaps an older sibling, which was temporarily looking after the infants, who were likely from multiple clutches.

The discovery hints at post-hatching cooperation in *Psittacosaurus*, whereby the parents entrusted others to care for the young while they went about their day. Although numerous types of bird cooperate in this way to

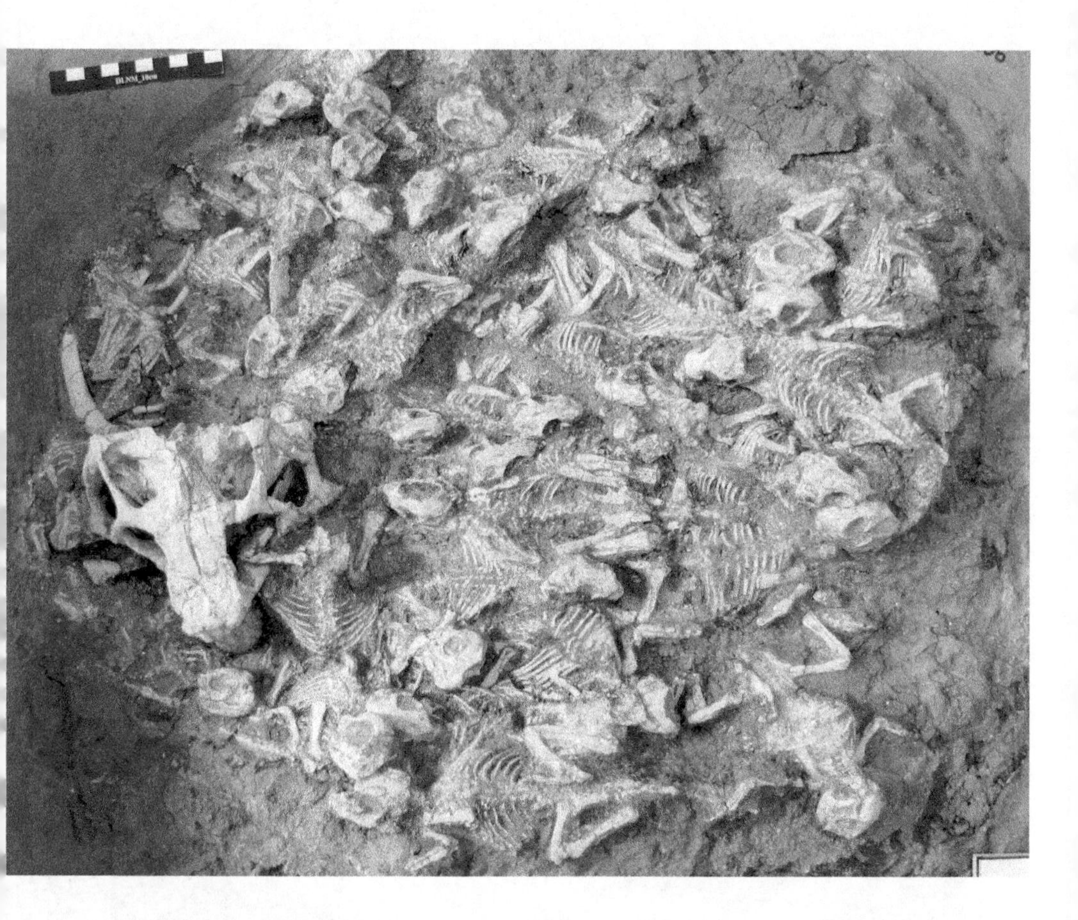

FIGURE 2.9. The subadult *Psittacosaurus* "babysitter," identified by the large skull on the left, preserved alongside the complete skeletons of numerous baby *Psittacosaurus*.

(Courtesy of Brandon Hedrick)

BOB NICHOLLS 2020

help rear their young, meerkats are one of the best modern analogues. Living in large colonies, watching over the young pups at the burrow, meerkat babysitters must ensure that the pups are kept out of danger from predators (such as lions, hyenas, and eagles) and bad weather. It is no easy task. Over the course of a twenty-four-hour shift, the average babysitter loses 1.3 percent of its body weight compared with those meerkats who continue foraging. Interestingly, the breeding pair never babysit their pups. Of course, comparing specifics between meerkats (a mammal) and *Psittacosaurus* (a reptile) is quite a stretch, but it nonetheless gives an idea of the effort and energy costs of cooperative rearing of young.

Additional assemblages of different-aged *Psittacosaurus* juveniles have also been reported and present further evidence of sociality. The babysitter fossil not only records an unusual behavior never before documented in a dinosaur, but it and the other *Psittacosaurus* aggregations show that this species engaged in complex social behaviors, offering a more complete picture of family life.

FIGURE 2.10. (Opposite). *Stay close to the big one.*

Leading the way through a dense woodland, a babysitter *Psittacosaurus* walks through the rain as the youngsters carefully follow.

DINOSAUR DEATH TRAP

The deadly, pack-hunting *Velociraptor* wowed audiences the world over with the release of *Jurassic Park* in 1993. For paleontologists, that year it was the announcement of a giant *Velociraptor* relative that stole the show—*Utahraptor*. At an estimated 7 meters long, this dromaeosaur ("raptor") was larger than *Jurassic Park*'s exaggerated *Velociraptor* and dwarfed the real version (*Velociraptor* was actually turkey-sized).

Beyond a few fragments of a skull, isolated bones, and claws, not much was known of *Utahraptor*, and it had remained poorly understood due to the lack of complete fossils. However, the recent discovery of a nine-ton sandstone block containing tightly packed *Utahraptor* bones is proving to be one of the greatest dinosaur finds of all time. It has shed light not only on this dinosaur's full appearance but also on how it might have interacted with other members of the same species.

"Did *Velociraptor* really hunt in packs?" is a question that almost every paleontologist has been asked on multiple occasions. Since *Jurassic Park*'s portrayal, this social behavior has remained ever present in the public's imagination. And yet, solid evidence for pack hunting or any social interactions in dromaeosaurs is especially rare.

The idea of pack-hunting behavior stems from the discovery of several specimens of the popular North American dromaeosaur *Deinonychus* (on which the *Jurassic Park* "raptors" were really based), found alongside a single large herbivorous dinosaur called *Tenontosaurus*. It was this association that sparked the idea that *Deinonychus* worked together to bring down much larger prey. Inevitably, this has been subjected to alternative interpretations—namely, that individual *Deinonychus* simply might have been scavenging a carcass or that perhaps the animals were swept up by water and buried together, though other *Deinonychus*–*Tenontosaurus* associations have been found, suggesting a likely relationship.

Direct evidence of social interaction came in 2007 after the discovery of multiple two-toed trackways found in Shandong Province, China. Dromaeosaurs stood on their third and fourth toes, whereas their second toe, which possessed the famous sickle-shaped "killing claw," was held aloft from the ground in a retracted position. In one specimen, six parallel and closely spaced trackways pointing in the same direction were found together, with the largest individual footprint being 28.5 centimeters long. This indicates that a small group of large dromaeosaurs were walking side by side and suggests some form of gregarious pack or family behavior.

Enter the new *Utahraptor* find. In 2001, geology student Matt Stikes happened on a large block with what looked like a human arm bone sticking out of it at a site near the town of Moab in east-central Utah, north of Arches National Park. The site came to be known as the Stikes Quarry. The find was reported to paleontologists, who examined the site and identified the bone as part of a dinosaur foot. After recognizing additional bones, they realized that this find was going to be something very special. Before the team could get too excited about the dinosaur treasure that might lie hidden inside the rock, they had to deal with the tricky situation of excavating and removing this enormous block from near the top of an extensive ridge (now called "*Utahraptor* ridge") without destroying any of the fragile bones.

After more than a decade of field excavations, led by paleontologists from the Utah Geological Survey, including Jim Kirkland, Utah's state paleontologist and one of the scientists who originally named *Utahraptor* in 1993, the nine-ton block was finally removed—in one piece—from the field and transported to the Thanksgiving Point Museum of Ancient Life in 2014. It has since been moved to the Utah Geological Survey's Research Center in Salt Lake City. Although the block is many years from being fully prepared and its complete story told, careful preparation of the 125-million-year-old Cretaceous rock has already revealed a mass dinosaur death assemblage.

Contained inside this dense block are the remains of multiple *Utahraptor* skulls and skeletons and at least two *Tenontosaurus*-like herbivorous dinosaurs. This is obviously a remarkable find, but what makes it so incredible is that the *Utahraptor* remains are of varying ages. Based on assessment of the bones documented so far, amazingly, there is one large adult, a subadult, five roughly two-year-olds, and three very young individuals estimated to be less than a year old. This provides paleontologists with ample

FIGURE 2.11. Out in the wilderness—the Stikes Quarry discovery site near Arches National Park, Utah, with close-up of the 9-ton block containing the *Utahraptor* skeletons, which was excavated and removed in one piece.

(Courtesy of Jim Kirkland)

FIGURE 2.12. (*A*) The tiny premaxilla (tip of the snout) of a baby *Utahraptor* found at the Stikes Quarry. (*B*) An adult *Utahraptor* skeleton.

([*A*] Courtesy of Scott Madsen; [*B*] courtesy of Gaston Design Inc.)

material to study the growth stages of *Utahraptor*, particularly how fast it grew and how it changed as it aged, but also to assess their relationship and whether this is a family group. Moreover, this only scratches the surface of the immense block, which appears to contain several times more specimens hidden deeper inside.

The presence of multiple *Utahraptor* individuals representing a growth series, and particularly the identification of several chicken-sized babies, strongly suggests that this was—in part, at least—a family. However, the geology of the site and the sandstone block in which the near-pristine bones are contained shows that this group had a particularly gruesome, sticky ending. They were trapped in quicksand.

This epic fossil appears to represent a predator trap. Seemingly, the herbivorous dinosaurs were the first to get stuck in the quicksand, and, attracted by the prospect of an easy meal, one by one the *Utahraptor* group found themselves mired as well. Although the family scenario seems most likely, that *Utahraptor* cared for its young and lived in groups (or packs), another interpretation is that the association might represent a combination of a family and a series of lone individuals. Whatever the exact scenario of their demise, all of these unlucky raptors were trapped, killed, and subsequently buried, the first evidence of multiple dinosaurs having fallen victim to quicksand.

This exquisite fossil might prove to be the best example of evidence for pack hunting and/or group sociality and parental care in dromaeosaurs. Nonetheless, with many years of skilled and delicate preparation work remaining and funding still required, the full story of this fossil will not be revealed until the entire specimen has been released from its rocky tomb.

FIGURE 2.13. (Overleaf). *The mire siren.*

A family of *Utahraptor* are attracted by the foul stench of a decaying dinosaur, but those that are too eager to feast on the flesh have become trapped in quicksand with no escape.

A PREHISTORIC POMPEII: AN ECOSYSTEM CAUGHT IN TIME

Yellowstone National Park is one of the most beautiful places in the world. The vast majority of this more than 3,000-square-mile wonderland is located in Wyoming and is famed for its incredible geysers, breathtaking scenery, and remarkable and diverse wildlife, which attracts millions of visitors each year. Yet, many people are unaware that Yellowstone is actually the site of a supervolcano—a volcano that erupts on a massive scale of more than 240 cubic miles. Technically it is called a *caldera*, which is the result of a volcano having essentially collapsed in on itself; its last major eruption occurred more than 600,000 years ago.

Reshaping the land and destroying wildlife, these supervolcanoes leave devastation in their wake. We can see the direct result of one of these epic eruptions by going a little further back in time, to around 12 million years ago. In this unique setting, hundreds of animals that lived in and around a popular watering hole were killed and buried by several meters of thick volcanic ash from the devastating eruption of a supervolcano.

Situated in Nebraska, this world-famous site is now a National Natural Landmark, appropriately known as the Ashfall Fossil Beds. The volcanic explosion originated about a thousand miles away in what is now southwestern Idaho, from the Bruneau-Jarbidge caldera. This caldera is part of what is known as the Yellowstone hotspot, a volcanic hotspot that over millions of years, through movement of the North American tectonic plate, now lies beneath the Yellowstone caldera. The resulting deadly ash cloud drifted eastward across the Great Plains, where it began to settle. A mass-death assemblage of a once thriving savanna-like ecosystem of animals was left behind, preserved in the very pose they died in—a sort of prehistoric Pompeii.

The first real clues that a catastrophic natural disaster had occurred were uncovered in 1971. While walking on farmland with his geologist wife, Jane, paleontologist Mike Voorhies found a fossilized intact baby rhinoceros skull embedded in a layer of ash. This was a life-changing discovery. He realized the potential significance of the find and that it could lead to more fossils—and he was correct. It took six years before any major excavations began, but it was worth the wait. The skull turned out to be articulated

FIGURE 2.14. (*A*) Multiple complete skeletons of *Teleoceras major* found in their original death pose, having suffocated from deadly volcanic ash, in which they remain buried. (*B*) Generic shot showing the extent of the site and various skeletons.

(Courtesy of Lee Hall)

with an entire skeleton and associated with many other complete skeletons. Here, a whole variety of animals, including early rhinoceroses, camels, and horses—entire herds and families—perished as a result of ash inhalation. To allow others to appreciate, learn, and understand what had happened, many of the three-dimensional skeletons were left in the ground in the exact positions they were found, and a building was erected over them. To this day, the site continues to be excavated and new discoveries made.

Volcanic ash contains microscopic particles of glass. Just imagine what inhaling this would do to your lungs, not just for a few days but for weeks. After inhaling the ash, these prehistoric animals became severely sick, and one by one, they began to die. In fact, the lung damage was so extreme in the horses, camels, and rhinos that abnormal bony growths formed on the skeletons, indicative of lung failure caused by a respiratory condition called Marie's disease (or *hypertrophic osteopathy*).

Small-bodied animals, such as turtles, birds, and musk deer, are buried close to the bottom of the 3-meter-thick volcanic ash layer, indicating that they were the first to fall victim; some even might have been buried alive. Their smaller lung capacities were unable to deal with the ash for any length of time, and they ultimately suffocated soon after it fell. The medium-bodied animals, such as multiple species of ancestral horse and camel, survived for a short while, at least a couple of weeks, but as they continued to inhale the poisonous gas, they, too, succumbed to a slow and painful death. Bones of some of these animals also bear scavenging marks, suggesting that carnivores such as bone-crushing dogs, whose remains are rare but have also been found, were actively feeding on the dead.

The last of the animals to die were the largest in the ecosystem: the rhinoceroses. Unlike rhinos today, which are typically solitary animals that sometimes form small groups (called *crashes*), a herd of more than one

FIGURE 2.15. (Overleaf). *Unaware of their doom.*

Going about their daily lives, an entire community of animals—including part of a herd of barrel-bodied rhino, camels, horses, and turtles—surround a large pond as a bone-crushing dog watches on. Death looms as an enormous plume of volcanic ash can be seen in the distance.

hundred mostly intact skeletons belonging to the species *Teleoceras major* were found together. Hippo-like in appearance, this barrel-bodied and relatively short-legged rhino spent part of its time in the water. Its skull was very large, especially in males, and it possessed a single small nose horn. Most of the skeletons, ranging from young calves to elderly adults, were females and young, with fewer older males represented. Some of the mothers lay buried with young calves nestled alongside them, others died with their last meal (grass) still in their mouths, and some left their final footprints in the ash. Remarkably, one pregnant rhino was even found with an unborn calf in the birth canal, which suggests that she might have been in mid-labor when she died. This specimen, coupled with the discovery of several young calves, suggests that breeding was seasonal, at specific times in the year, and therefore similar to living mammals such as bison.

Based on the age, sex, and number of the rhinos, there is clear evidence of social structure in this prehistoric herd, which shows that these animals both lived and died together, although whether they lived together year round is not possible to say. Young adult males are absent, and older males are greatly outnumbered by females, which suggests that the herd might well have had a polygynous breeding system, whereby several dominant males mated with multiple females.

Just take a moment to consider this exceptional story. A disastrous and deadly supervolcano led to the fine preservation of one of the most amazing mass fossil sites in the world. Ashfall provides a unique glimpse of a once-thriving ecosystem of animals that were reluctant to leave their local waterhole, but which ultimately led to their suffocating and being trapped for all eternity, only for their remains to be discovered by chance millions of years later.

FISH TRAPPED INSIDE GIANT CLAMS

In the warm waters of the South Pacific and Indian oceans live the largest mollusk on the planet, the giant clam. Weighing more than 200 kilos and reaching up to an impressive meter or so in length, these are massive mollusks. They live in species-rich coral reefs, and their large shells are used as shelters by various animals and act as a form of protection against predators; they make excellent nurseries for fish. This is a type of symbiotic relationship known as *commensalism*, or, more specifically, *inquilinism*, whereby two species interact, with one reaping all the benefits but without causing any harm or hindrance to the other.

Many types of bivalve play hosts to various fish. Commensalism is widespread throughout the animal kingdom, and there is every reason to presume that similar relationships must have been in place in prehistoric ecosystems. However, finding conclusive evidence of such interactions can be problematic. Different fossil species are often found together, but careful interpretation is required to determine if any specific, potentially interacting relationships are present rather than just random associations.

Inoceramids are an extinct family of bivalve mollusks that are especially common, have a global distribution, and include the largest known bivalve ever found, *Platyceramus platinus*, a species that could reach nearly three meters in length. Complete specimens, though common, are normally flat and very thin, which means that they are often easily damaged and fragmented, making them difficult to collect.

One 85-million-year-old *Platyceramus platinus* collected from Cretaceous chalk deposits, known as the Smoky Hill Chalk in Kansas, was reported in 1929; it was found to contain three specimens of a new species of fish. It was not until the 1960s that this association was recognized as unusual. Paleontologists realized that certain species of fossil fish in these deposits were commonly found inside the shells of inoceramids, and they pondered why. The more time they spent looking for these bivalve–fish associations, the more specimens were found, which led to the discovery of several new fish species. An ingenious way of fossil fishing.

Quite remarkably, more than 100 inoceramids from the Smoky Hill Chalk have been found to contain fish, and, astonishingly, altogether more than 1,200 individual fish are preserved inside these bivalves. Additional specimens are recorded from Colorado. Having numerous examples of the same association is strong evidence that these specimens represent actual interactions between different species rather than chance happenings.

As many as nine fish species, including small members related to bowfin fish, squirrelfish, beardfish, and one eel, have been found inside the enormous shells of *P. platinus* and other bivalves. Most shells contain numerous

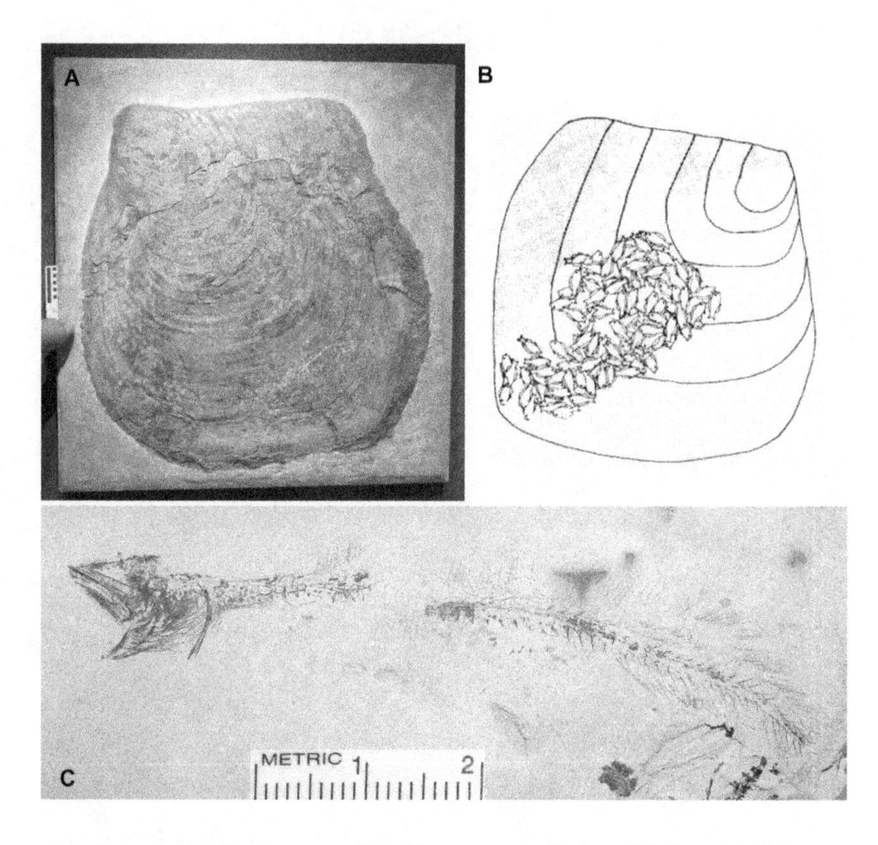

FIGURE 2.16. (*A*) A giant, complete *Platyceramus platinus*. (*B*) Illustration showing multiple fish found inside a *P. platinus* shell. (*C*) Finely preserved *Urenchelys abditus* eel, an example of a fish found inside one of the giant clams.

([*A*] Photograph by the author; [*B*] slightly modified from Stewart, J. D. 1990. "Niobrara Formation Symbiotic Fish in Inoceramid Bivalves." In *Society of Vertebrate Paleontology Niobrara Chalk Excursion Guidebook*, ed. S. Christopher Bennett, 31–41. Lawrence, KS: Museum of Natural History and the Kansas Geological Survey; [*C*] courtesy of Mike Everhart)

BOB NICHOLLS 2020

individuals of the same type of fish, but others include multiple species, suggesting that they also interacted with one another as well as with the bivalve. Extraordinary are those with an abundance of individuals inside; one contained a whopping 104 tiny fish. These examples are typical of shoals, where all the fish are of a similar length and belong to the same species.

These life associations represent commensal relationships between *P. platinus* and various fish species. It was found that some of the shells retained evidence of their gills, indicating that they were alive when the fish interacted with them, whereas others did not have any gill remains. The fish might have resided within both living and dead gaping shells, using them as shelters to hide from predators and presumably for other reasons (perhaps mating and feeding). Why, then, have so many fish been found fossilized inside the shells?

Clearly, the relationship was advantageous for the fish, but the association had deadly consequences. Not because the inoceramids lured the fish and fed on them, but if their shell rapidly closed, perhaps on death or when the muscles and ligaments holding the shell agape had been eaten, anything that was inside became trapped. The fish were therefore highly likely to have been alive when entombed with no form of escape. Many of the fish are often located in the most domed part of the shell, and this might indicate the position of the last remaining oxygen inside the shell, where the fish congregated. As with many modern-day reef fish that are found inside the shells of giant clams, a similar, albeit deadly, relationship was just as common in this prehistoric ecosystem.

FIGURE 2.17. (Opposite). *Our refuge and tomb.*

Numerous species of fish related to bowfin fish, squirrelfish, beardfish, and one eel are living in and around the enormous shells of *Platyceramus platinus*.

SNOWMASTODON: A REFUGE FOR SMALL ANIMALS

As is typical in paleontology, many great fossil finds are made by chance, sometimes in unusual places or odd circumstances. Ongoing construction work to enlarge a reservoir near the picturesque village of Snowmass, a ski resort complex in the beautiful Colorado Rockies, led to the discovery of the most significant high-elevation ice age fossil site in the world.

The Ziegler Reservoir fossil site, or simply Snowmastodon, dates to the Pleistocene epoch, between 55,000 to 140,000 years ago. During this time, the site was initially a conifer forest surrounding a warm lake, which gradually became a cooler and drier marsh or wetland area, an ideal habitat that attracted lots of ancient animals, who lived and died in the area.

On October 14, 2010, bulldozer operator Jesse Steele made a once-in-a-lifetime discovery. He was routinely pushing soil with his dozer when suddenly his blade flipped over two giant ribs. Inspecting the bones, he found several more lying in the soil. Amazingly, he had run over the skeleton of a Columbian mammoth. These first few finds sparked interest from paleontologists at the Denver Museum of Nature and Science, who were given permission to lead a large-scale excavation. As winter rapidly approached, poor weather conditions and an agreed deadline to allow construction work to resume by July 1, 2011, meant that the team had to race against time to collect as many fossils as possible. Finds like this do not come along very often, so paleontologists have to take advantage of every opportunity.

The dig began almost immediately, and fossilized bones were found scattered everywhere. But heavy snow and frozen ground halted the team's progress, and their excitement had to be curtailed for a tormenting six months. Having to remain patient like this, knowing that there are so many fossils simply waiting to be collected, is one of the most frustrating things for a paleontologist. When the dig resumed, they had just seven weeks to complete the excavation.

Several hundred people, including scientists, volunteers, and members of the construction crew, came together in an enormous team and took part in what became the biggest fossil dig in Colorado history. This was quite an accomplishment, considering that Colorado has such a rich fossil heritage, being one of the greatest places in the world to find Jurassic dinosaurs.

Overall, more than 7,000 yards of dirt were removed, mostly by hand. More than 30,000 bones were collected, along with well-preserved plants and invertebrates. The impressive megafauna included extinct species of camel, bison, and giant ground sloth, once native to Colorado, alongside fossils of deer, black bear, coyote, and bighorn sheep, which thrive in the same area today.

Of all of the fossils, the largest animals were the mighty mastodons and mammoths, cousins of today's elephants. Most of the mammal remains belong to the American mastodon, *Mammut americanum*, with at least 35 individuals collected—a testament to the importance of the discovery, given that previously only three mastodons had ever been found in the whole of Colorado. Bones of at least four partial skeletons of the slightly larger Columbian mammoth, *Mammuthus columbi*, were also found. These were among the largest terrestrial animals during the Pleistocene epoch, and just one dead giant would have fed numerous carnivores, big and small, for several weeks. But what of the remaining skeleton? Well, this provided shelter and a home for many of the smaller animals that lived and died alongside them.

While admittedly impressive, the megafauna were not the only major talking point at Snowmass. A whole variety of smaller animal communities were also found. Fossils of small-bodied mammals still native to the area were collected as part of the dig, such as shrews, beavers, squirrels, and chipmunks, along with various birds, reptiles, and amphibians. By far the most common of the small animals was the tiger salamander, *Ambystoma tigrinum*, with more than 22,000 bones recorded. The same species of salamander still exists in North America, where they are commonly kept as pets; it is the only native salamander in Colorado. As the excavation came to an end, it seemed that was the last of the new discoveries, but fossils have a tendency to slowly reveal their secrets.

As the freshly collected bones were cleaned, many tiny and delicate remains of various small animals were found buried deep inside them. This was a fortuitous and unexpected surprise. During the preparation of a mastodon, one particularly good and partially complete tiger salamander skeleton was revealed inside the pulp cavity of the mastodon's tusk. Had the salamander crawled inside for temporary shelter, perhaps for protection

from a predator or adverse weather, or was it resting inside its "home"? Several other mastodon tusks were also discovered to be jam-packed with tiny salamander bones. As the same species still survives and thrives today, this can help to shed some light on what its prehistoric relatives were doing.

Tiger salamanders live in underground burrows, under rocks, and inside logs near ponds, lakes, and streams. Considering that the mastodons had died near or in water, their giant log-like bones and tusks would have seemingly presented a perfect shelter in an ideal environment. Clearly, as reflected by the huge number of bones, conditions at prehistoric Snowmass suited the salamanders. The many small, contemporaneous animal remains found inside the bones and tusks present compelling evidence that they were also living in and around the large skeletons.

This unusual and unique association provides solid evidence of ecology, showing how these animals were interacting with each other and their environment. It is easy to imagine a salamander escaping the baking heat of the sun by hiding inside a tusk, or squirrels and chipmunks running up and down, in and out of the bones. It is the sort of story that seems plausible but one that you would expect could not be interpreted from fossils—and yet, here we are. There is no doubt that this is a snapshot of one animal community interacting with another across the distance of many thousands of years, breathing new life into old, dead giants.

FIGURE 2.18. (Opposite). *The Snowmastodon community.*

A wide variety of small animals living in and around the long-dead skeletons of two American mastodons, *Mammut americanum*. Pictured is a crane, a finch, a lizard, a family of chipmunks, a grey squirrel, a mouse, two voles, a snake, tiger salamanders, and hundreds of caddisflies.

GIANT FLOATING ECOSYSTEM: A JURASSIC MEGARAFT CRINOID COLONY

When walking through a museum and admiring the displays, occasionally you see an object that takes your breath away, that one thing you know will forever stay in your memory. As a paleontologist, I get to see some pretty incredible fossils, but one of the most visually spectacular fossils I have ever seen is the world's largest colony of sea lilies, a Jurassic giant extending more than one hundred square meters.

Commonly known as sea lilies and feather stars, crinoids are often mistaken for underwater plants. However, they are a type of echinoderm, the same animal group that includes sea urchins and starfish. These marine filter-feeders first appeared in the fossil record more than 450 million years ago, and around 600 species live in the world's oceans today in both shallow and deep water.

FIGURE 2.19. The Hauff specimen of the gigantic floating crinoid *Seirocrinus subangularis* with close-up of the exceptional details, as displayed at the Hauff Museum, Holzmaden, Germany.

(Photographs by the author.)

In March 2017, as part of a research trip in southern Germany, I accompanied my friend and fellow paleontologist, Sven Sachs, to the Urweltmuseum Hauff (The Hauff Museum of the Prehistoric World). Situated in the small town of Holzmaden, a town world-famous for its Early Jurassic fossils, the museum contains the best collection of specimens from the area, collected by four generations of the Hauff family. Continuing the legacy, collecting, preparing, and studying fossils, is Rolf Bernhard Hauff, the current museum director. It was during my visit to the museum that I met Rolf, who gave me a tour and introduced me to the enormous crinoid fossil, which was found in 1908 and took eighteen years to prepare.

Like a beautiful piece of natural artwork, this giant colony comprises several hundred small and large individual crinoids, some of them exceeding an impressive 20 meters in length, with crowns a meter in diameter. They all belong to a single species called *Seirocrinus subangularis*, which had a long, ropelike stem with a large, cuplike crown and numerous feathery arms that were used to trap tiny bits of food from the water. Rather than being attached to the seafloor, these crinoids are anchored to a giant, 12-meter-long tree trunk. With the crinoids fixed mostly to the underside of this giant floating log, alongside countless encrusting oysters, the driftwood acted as an enormous substrate for the creatures. Hanging upside down in the water, they took advantage of the association, having a free ticket to ride across the warm tropical seas and develop into immense filter-feeding communities of adults that were capable of reproducing.

Once attached to a substrate, *Seirocrinus* became immobile. Therefore, young, free-floating crinoid larvae must have first settled on the giant logs before maturing to adulthood and colonizing them. Many of the crinoids attached to the log are adults, so comparing their growth rates with rates of living relict relatives of *Seirocrinus* suggests that these megaraft colonies might have existed for more than ten years and perhaps up to as many as twenty years. This exceeds the life expectancy of equivalent modern-day raft systems.

FIGURE 2.20. (Overleaf). *The transient oasis.*

An enormous *Seirocrinus subangularis* barge floats across the ocean and creates a habitat for a community of animals, including ammonites, belemnites, and fish. Two curious ichthyosaurs (*Suevoleviathan*) inspect the megaraft.

Bob Nicholls 2020

The log decayed over time and was infiltrated by water. That, combined with the total weight of the crinoids and other attached animals (such as oysters), caused the log to finally sink. The crinoids and associated fauna died and were buried in the anoxic seabed, where they were perfectly preserved. Such fossil colonies are among the largest in situ (still preserved in their original setting) invertebrate accumulations ever to be found in the fossil record.

Given the unlikely circumstances for the preservation of this drifting colony, it might be surprising to read that this is not a one-off discovery. Quite the opposite. Exceptional colonies like this are known from around 100 specimens of varying size (although not as big as the Hauff specimen), and practically all are from Holzmaden and surrounding areas; other crinoid species have also been discovered attached to logs. Specimens have been found around the globe, showing that this special adaptation allowed these crinoids to travel far and wide across the ancient oceans.

Similar analogies can be shown with modern-day driftwood communities. When trees are washed out to the ocean, they can become important habitats for scores of plants and animals, among them bivalves, sea anemones, and crustaceans such as barnacles. These floating rafts also attract fish, sea turtles, and seabirds, who feed on the associated organisms.

Just like their present-day analogues, these giant, ancient floating megarafts not only acted as drifting homes for the crinoids, but their great size seemingly would have provided excellent cover for a diverse animal ecosystem. Although not found in direct association, a plethora of marine animals are preserved in the same rocks as these crinoids, including squid-like ammonites and belemnites, fish, and marine reptiles (such as ichthyosaurs), and presumably they interacted with these rafts in a fashion similar to species interactions today. These 180-million-year-old fossil colonies provide a snapshot of an unusual ancient ecological community.

3 Moving and Making Homes

The Great Wildebeest Migration is one of nature's most phenomenal events. A thing of beauty, this spectacular annual migration sees over one million wildebeest and thousands of other animals migrate over a year-long trek between Kenya and Tanzania. It is the largest terrestrial mammal movement on Earth. The 1,000-kilometer round trip is, however, filled with danger as the herd is forced to cross fast-flowing, crocodile-infested rivers and evade one of the world's largest populations of lions, all while caring for newborn calves. Falling afoul of predators, disease, hunger, thirst, and, ultimately, exhaustion, several hundred thousand members of the horde do not complete the migration.

Travelling far and wide in search of their favorite feeding grounds and exploring new locations along the way, these mammals find mates, pass on their genes, and gain an advantage over their competitors. Such benefits far outweigh the inevitable deadly risks. But whatever the case for migration, be it wildebeest and company embarking on their journey across the land, birds flying vast, record-breaking distances, or whales swimming thousands of miles, at heart this behavior is all about survival.

Species differ substantially in what makes them get up and go, whether caused by seasonal changes in the weather; the need to procure food, find mates, or escape; or simply to stretch their legs (if they have legs). In such scenarios, animals often leave evidence of their movement, most obviously in the form of tracks. By looking at the tracks' shape and size, among many other things, it is possible to determine which animal left its tracks behind, whether it was traveling alone or as part of a group or herd, how fast it was moving, or even if the animal was on the hunt or being hunted.

Similarly, an animal's home can tell you a lot about its life. Living inside a home, either permanently or temporarily, is a major adaptation that many animals have evolved to enhance their survival. A home can be as simple as a hole in the ground or a suitable space inside a cave or tree, or it can be more complex, like deep and extensive burrows or intricate nests. Such shelters provide animals with protection from predators and extreme weather, a place to rear young and store food, and an opportunity to relax.

Carefully selecting where and when to create your home, whether journeying to strange and faraway lands in search of the ideal location or simply identifying the best tree or rock crevice in the vicinity in which to settle down, is one of the most important decisions these animal architects must make. Sometimes, however, animals living in custom-built homes are not always the creators. Many species patiently wait until the hard work is done and then move in as a lodger or, worse, claim the home for themselves, forcing out the occupants.

The home might be intended to last a lifetime or just long enough to serve a specific purpose. Tiny termites, for example, are among the most dedicated builders. Some species form underground nests with interconnecting chambers and tunnels and build giant mansion-like mud mounds that can tower more than 10 meters above the ground. Enormous colonies live inside these homes, which can take years to complete.

On the other hand, temporary homes tend to serve one primary purpose; for instance, the maternity den of a pregnant polar bear. During the winter months, instead of hunting for food, the female digs an igloo-like snow cave that acts as a temporary shelter. Shielded from the elements and hidden from predators, she safely gives birth and nurtures her cubs while all are at their most vulnerable. Although great care and effort can go into creating such dwellings, in many circumstances, temporary shelters are basic and require little or no effort to construct. It can be as simple as a mouse sheltering under a log or a rock while waiting for the rain to stop, or something bizarre, like the empty shells of mollusks (or even coconuts) that an octopus carries to conceal itself to avoid detection from predators or prey.

Studying such tracks and traces of a living animal helps in gaining a more complete picture of an individual's life and how it interacts with the habitat in which it lives. It is possible to follow a track and find the animal at the end of it or look inside a burrow and see what it contains. Most significantly, we can watch animals leave their traces in real time. Therefore, identifying the trace-maker can generally be straightforward. Unfortunately, it is not quite as easy when it comes to fossils, although not nearly as frightening as discovering a living bear inside its den!

Over its lifetime, an animal might leave behind countless footprints, several homes, or other traces but only one skeleton. Arthropods are the exception. They leave behind multiple molts, although just one carcass (the actual dead body). Apply this logic to fossils, and it becomes easier to understand why, in many instances, trace fossils are more common than body fossils. In most cases, it is possible to say which type of animal left the trace, but it is almost impossible to say exactly which species made it. Only in rare circumstances are both trace and body left behind by the same individual and preserved together.

Unravelling how long-extinct animals moved and made their homes can provide all-important clues for uncovering events that happened thousands to millions of years ago. However, major barriers exist. Dealing with fossils, we cannot watch these ancient animals move or build homes. So, do we have enough reliable evidence of such behaviors, and are we able to apply the same methods that we use when studying and interpreting living species? In theory, yes, but it can be very challenging.

Fossilized tracks and burrows provide some of the best direct evidence for understanding the behaviors of prehistoric animals in general. The study of tracks and other such trace fossils is called *ichnology*. Using a trackway as an example, by carefully studying the track shape and type and comparing it with the foot anatomy and size of animals found in the same aged rocks or prehistoric habitat, we can deduce which type of animal created the tracks.

It would, however, appear almost impossible to determine exactly which species made such traces. Yet, we can go a step further.

Would you believe me if I told you that a multi-million-year-old burrow with its creator still inside has been found or that we have evidence of prehistoric animal migrations? As unlikely as it might seem, such remarkable fossils *do* exist. In this chapter we will dig deeper into the behaviors of these prehistoric movers and homebuilders.

MAMMAL MOVERS: TRAGEDY AT THE RIVER CROSSING

Understanding the behaviors of living mammal herds, and having an appreciation for the obstacles they must overcome, can inform us greatly about their prehistoric counterparts. Having been explored in detail and studied extensively, many North American fossil-bearing locations have produced some of the most spectacular fossil gatherings. In one exceptional circumstance, while fossil hunting in southern Wyoming in 1970, a team from Chicago's Field Museum made a startling discovery. They came across a graveyard of at least twenty-five rhino-like brontotheres buried in an area less than 100 meters square, which was excavated over three sunny summers.

Brontotheres belong to an extinct family related to rhinos, horses, and tapirs. Being large and bulky, they outwardly resemble rhinos, and some species have elaborate horns. Their fossils are particularly well known across western North America and central Asia. All the skeletons discovered in Wyoming belonged to a type of horse-sized, hornless brontothere called *Metarhinus*. Studies of their teeth showed that the group contained a mix of ages, from very young individuals—some no older than a couple of months—to senior adults of both sexes.

This monospecific group (a group containing only animals of the same species) most probably represents a small fraction of a once giant brontothere herd. Given that the bones were found in close association and in exactly the same layer of rock, it is clear that they perished together. But the mystery of what happened to them is a secret locked inside the rocks in which they are buried.

The geology of the graveyard shows that the animals were preserved in a prehistoric floodplain deposit resulting from a single flash flood event, pointing to a tragic accident leading to the loss of life. Interestingly, no other large or small mammals or any other vertebrates were found in the same layer as the specimens, which lends further support that this was an instantaneous mass death event of many individuals from a single herd.

Heavy rains resulted in a river bursting its banks and becoming flooded. A brontothere herd was challenged with overcoming a treacherous obstacle that would prove to be too hard a task for some. Mammal herds routinely cross rivers, often with no cause for concern, but depending on the severity of the conditions, sometimes members of the group can find themselves in grave difficulty. As is typical when large herds are crossing rivers—especially flooded or fast-flowing ones—animals tend to trip, become stuck in the mud, and get trampled, or they can be pushed underwater and even misjudge the height of the river banks on the other side. Naturally, in such circumstances, the inexperienced or weak and elderly are more prone to making these deadly mistakes, which coincides with the age of several of the unfortunate members of this brontothere herd.

Whatever led them to cross this river and ultimately die, the unlucky members of the herd were engulfed by the water, only for their bodies to be swept downstream, where they accumulated as a carcass jam. It is therefore probable that many, if not all, of these individuals drowned and were already dead before they ended up in their final resting place. Although these animals are separated from their contemporaries by 40 million years, there is an obvious parallel with the accumulations of wildebeest that suffer the same consequences when attempting to cross the Mara River during the Great Migration.

This scene of mass mortality captures a specific moment when a small part of a once large brontothere horde perished during a dramatic prehistoric flash flood. Herds form for several reasons, protection being among the most important, because there is safety in numbers. Social behavior differs from one group to the next, but, as in many mammal herds today, this prehistoric herd composed of animals of different ages and sex and shows that—for some time at least—they stayed with and cared for their young.

This discovery is among some of the most compelling evidence of sociality in prehistoric mammal herds that has continued through the ages and is a major, fundamental part of herding behavior to this day.

FIGURE 3.1. (Opposite). *Tragedy at Thunder River.*

A large herd of *Metarhinus* brontotheres attempts to cross a fast-flowing flooded river, with deadly consequences.

BOB NICHOLLS 2020

FOLLOW THE LEADER: THE EARLIEST ANIMAL MIGRATIONS

One of the most instantly recognizable fossils, trilobites are a hugely diverse and captivating group of extinct marine arthropods whose remains have been found on every continent. With a rich record extending at least 521 million years into the past, they are among some of the earliest arthropods to have evolved. But they were wiped out a quarter of a billion years ago during Earth's most violent mass extinction of all, at the end of the Permian. Before dinosaurs even appeared, trilobites were already fossils under their feet.

They were just as common in the ancient primeval oceans as their crustacean cousins are today. More than 20,000 species have been recognized so far, but despite the wealth of fossils and extensive research on them, little direct evidence of their behavior has been captured. This is why the discovery of conga line–like "queuing" trilobites is so intriguing.

Representing the oldest known evidence of the blind leading the blind, albeit in the literal sense, multiple groups of an eyeless, three-spined trilobite called *Ampyx priscus* were collected from 480-million-year-old Ordovician rocks in southeastern Morocco, near the town of Zagora. Similar-aged queues formed by the same species have also been found in southern France.

The number of *Ampyx* preserved in these queues varies from three to twenty-two individuals, and all are arranged in the same direction, usually lined up one by one in a head-to-tail manner, often with their bodies and/or spines contacting or overlapping one another. They are generally complete, and none is disarticulated, representing actual carcasses rather than molted exoskeletons (or *exuviae*). These details suggest that the marching trilobites are preserved in the original positions in which they perished, indicating that the groups were rapidly buried, probably during storm events, either while alive or immediately after death.

This phenomenon is not restricted to one species. Similar distinct queues have been reported for several other trilobites, including additional specimens from Morocco but also from Poland, Portugal, and elsewhere. Notably, seventy-eight queues were collected from 365-million-year-old Devonian rocks in the Holy Cross Mountains of central Poland, each with up to nineteen aligned and articulated individuals. All of the trilobites belong to another blind species called *Trimerocephalus* and, like the *Ampyx* specimens, are arranged in a unidirectional manner, in contact with one another, and preserved in their original life positions.

What exactly caused these queues of death? Some have argued that the trilobites had formed queues inside burrows and were buried within them, whereas others have suggested that they were accumulated by ocean currents. Neither of these interpretations holds water. First, there is no evidence or any indications of the marching trilobites buried inside burrows or tunnels. Second, the fine preservation, contact between specimens, and their unidirectional nature rules out that they were randomly brought together by currents. So, what gives?

Similar queues are formed by various present-day arthropods when they migrate. The best example is the spiny lobster, which performs mass single-file migrations, possibly triggered by seasonal storms or for traveling to breeding grounds. Linked through contact between the tail fan of the preceding individual and the front appendages (such as the antennae) of the next, these lobsters form long chains and migrate over vast distances, often traveling day and night for several days. Moving in such chains decreases hydrodynamic drag, saves energy, and reduces the probability of predation. This seasonal migration behavior is a perfect fit for the trilobite queues, which were also likely to have formed due to environmental stresses (such as storms) and/or to reach their distant spawning sites. Following the leader, the blind trilobites would have relied on physical contact from their appendages, like the remarkably long projecting spines in *Ampyx*, but they probably also used chemical communication to locate, join, and follow individuals in these queues.

Trilobites are not the only ancient arthropods to be found in similar associations. The oldest unequivocal evidence of such collective behavior comes from the discovery of a strange, small shrimp-like arthropod from 520-million-year-old Cambrian rocks in Chengjiang, China. Numerous groups of these creatures, called *Synophalos*, were found interlocked in unidirectional chain-like queues of two to twenty individuals that are linked with their tail fan inserted into the carapace of the individual behind it. The name *Synophalos*, which means "to journey in company under the sea," was given in light of their association. It was first hypothesized that these

FIGURE 3.2. (*A*) Migrating *Ampyx priscus* trilobites preserved one after the other, collected from near Zagora, Morocco. [*B*] A migrating queue of living spiny lobsters in the Bahamas.

([*A*] Courtesy of Jean Vannier; [*B*] courtesy of Błażej Błażejowski)

Synophalos chains swam through the water column as one large unit, unlike anything that had been seen before. However, some have presumed that they were more like the spiny lobsters and trilobites and instead marched in unison on the sea floor. The latter assumption appears more likely. Just like *Ampyx* and *Trimerocephalus*, these early social arthropods were probably migrating and died en masse.

Over half a billion years ago, animals evolved the first brains and sensory organs. Taking advantage of those attributes, these early arthropods were the first animals to develop a complex form of collective behavior and coordinated migratory gatherings that enhanced their chances of survival and reproduction in a rapidly evolving world.

FIGURE 3.3. (Opposite). *We follow.*

A scurrying queue of migrating *Ampyx* trilobites following one by one across the ancient seafloor.

BOB NICHOLLS 2020

SITTING ON THE JURASSIC BAY

When you walk along a beach and leave your footprints in the sand, a trace of your behavior is momentarily captured in time. Just like us, dinosaurs left their tracks behind, too, but not all of them were washed away. Tracks are the most common trace fossils attributed to dinosaurs and are found globally.

We can discover a great deal about a dinosaur's behavior by studying its tracks. Did it zigzag across a beach, stop and change direction, or was it walking or running? Maybe it was being followed, or perhaps it was part of a herd. These types of scenarios offer some of the most visual stories, although occasionally the rarest of tracks result from something simple.

In the 1850s, a coffee-table-sized block of rock containing a beautiful pair of theropod footprints preserved with long metatarsal (heel) impressions was discovered in Massachusetts. The Early Jurassic fossil represents a resting trace, capturing a split second in time when this dinosaur squatted in the mud and rested. Initially, the trace was thought to have been made by a giant bird, which is unsurprising, considering that extinct theropod dinosaurs were bipedal and their footprints are strikingly similar to those made by living birds.

Curiously, on either side of the trace, the rock surface is covered in numerous minute spherical marks that were originally interpreted as raindrop impressions (which can and do fossilize), painting a rather somber moment of a dinosaur sitting on the beach waiting out a Jurassic rain shower. However, it has since been argued that the raindrops more likely represent gas bubbles escaping from the Jurassic mud as the theropod sat down. Rain or no rain, this fossil is impressive, but it is not alone.

More than ten dinosaur trace fossils showing a similar bird-like resting pose have been reported, each with a side-by-side pairing of the footprints and metatarsal impressions. The resting nature of the traces does not necessarily mean that their maker was in a state of inactivity; instead, it might suggest that the animal was taking time to clean itself, feed, drink, or something similar.

The most exceptional example comes from a similarly aged track site known as the St. George Dinosaur Discovery Site at Johnson Farm, in southwestern Utah, where multiple tracks were left by a variety of animals (mostly theropods) walking along a muddy beach around 198 million years ago. (Raindrop impressions have also been positively identified on the same track surface.) Cutting right across the site, one distinct trackway 22.3 meters long records the instant when a theropod simply stopped in its tracks and sat on a slope for an undetermined amount of time before standing up and walking away, left foot first.

FIGURE 3.4. The Early Jurassic theropod resting trace from Massachusetts, showing the perfectly preserved metatarsal impressions and the apple-sized "butt print."

(Reproduced from Hitchcock, E. 1858. *Ichnology of New England: A Report on the Sandstone of the Connecticut Valley, Especially Its Fossil Footmarks*. Boston: William White, 232.)

Fossil footprints are found to be only rarely associated with different traces reflecting other behaviors. In this case, not only did the theropod crouch down, shuffle into a comfortable position, and rest in a bird-like pose, but it left impressions of thickened skin from around the buttocks (the ischial callosity, or sitting pad), marks from the tail, and even handprints. Notably, the hand impressions provide details of their structure and anatomical position, showing that the palms faced each other (as if clapping) rather than facing down in the infamous "bunny hands" position of theropods in *Jurassic Park*.

Coincidentally, speaking of *Jurassic Park*, *Dilophosaurus* remains have been found in slightly younger rocks in the neighboring state of Arizona. Although it is impossible to say which species left the traces behind, the size of the footprints, handprints, and crouching position indicate that it was created by a medium-sized theropod (of about 5–7 meters in length), matching *Dilophosaurus* in size and rendering it a suitable stand-in model for the track maker. Oh, and if you were wondering, no acid was found. There is zero evidence that *Dilophosaurus* (or any dinosaur, for that matter) could spit acid.

Fossils like these go beyond your typical dinosaur tracks and traces and allow for a wealth of information to be extracted from a common and simple behavior like sitting and taking a rest, capturing a little moment in the day of a dinosaur.

FIGURE 3.5. (Opposite). *A place worth sitting.*

A *Dilophosaurus* rests on a muddy beach as rain begins to fall. Based on the theropod resting trace found at the St. George Dinosaur Discovery Site in Utah.

BOB NICHOLLS 2020

THE DEATH MARCH: THE FINAL STEPS OF A JURASSIC HORSESHOE CRAB

Picture an animal with a hard, dome-shaped shield that protects its body, a long pointy tail, ten legs, multiple eyes, and blue blood. It might sound like an alien from an upcoming movie, but this is a horseshoe crab. These archaic animals have changed very little in their overall body plan since the group's first appearance in the fossil record almost half a billion years ago. Their name is a little misleading, though, because they are not actually crabs but are, instead, more closely related to arachnids such as scorpions and spiders.

Countless Late Jurassic fossils, including horseshoe crabs, have been collected from various limestone quarries near the famous fossil grounds of Solnhofen, a village in Bavaria, southern Germany. The fossils are a by-product of extensive limestone quarrying that extends back to Roman times. Due to the many exceptionally preserved specimens discovered there, it is among the best places in the world to find fossils of this age and is on every paleontologist's bucket list of places to visit.

During the late Jurassic period, the Solnhofen area was a series of lagoons that formed part of an archipelago that teemed with life. Many lagoons contained high concentrations of salt washed in from the nearby warm, shallow sea, which resulted in their becoming stagnant and oxygen-free near the bottom. Death loomed for most animals unlucky enough to fall into these toxic bottom waters.

In 2002, an exceptional horseshoe crab fossil was discovered inside a quarry near the village of Wintershof, close to Solnhofen. Just imagine walking into a quarry and finding a 150-million-year-old horseshoe crab poking out of a limestone rock. As you investigate, you notice a series of marks in the limestone behind the horseshoe crab. Paying close attention, you follow these marks, carefully splitting the rock with your hammer and chisel, only to realize that these marks are tracks left by the horseshoe crab. This was the discovery of the longest fossilized death track in the world, in which the trace and its maker are preserved together in a final, epic death march—scientifically termed a *mortichnion*.

It is difficult to say for certain how the horseshoe crab came to be in this prehistoric predicament, although we have a pretty good idea. At

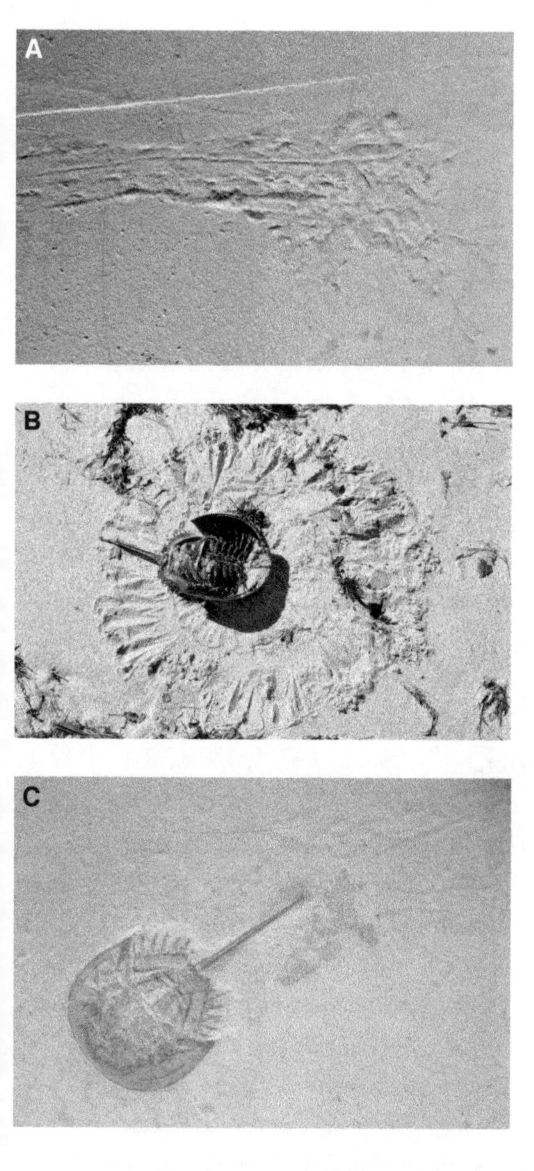

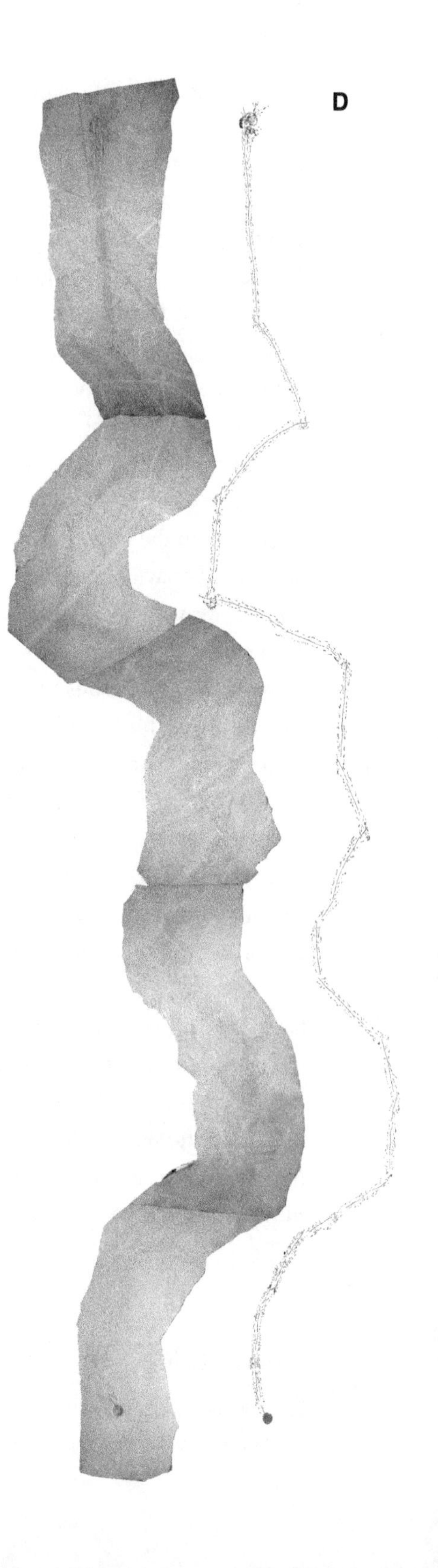

FIGURE 3.6. (*A*) The landing mark where the horseshoe crab first came to rest at the bottom of the lagoon; the circular depression was made when it landed on its back (on the prosoma). (*B*) Living horseshoe crab creating very similar depressions and markings in the sand, as inferred for the fossil. (*C*) The juvenile horseshoe crab that left the multi-meter track. (*D*) The entire death march with interpretive illustration.

([*A*, *C–D*] Photographs by the author; [*B*] courtesy of Sandy Tilton)

Solnhofen, animals were often washed into these lagoonal toxic traps during heavy, harsh storms or floods. Only a few of the hardiest animals, such as horseshoe crabs, were still alive when they reached the bottom, and thus, they left their tracks in the undisturbed soft mud. This particular horseshoe crab, of the species *Mesolimulus walchi*, was among the hardiest of them all, leaving behind a record-breaking 9.7-meter-long trackway (which I studied and formally described with fellow paleontologist Chris Racay while at the Wyoming Dinosaur Center). All the more impressive is that the culprit was a juvenile, just 12.7 centimeters long. In a manner comparable to living horseshoe crabs, when spooked, the young and inexperienced swim up into the water column, whereas adults usually do not. It is possible that such behavior made this Jurassic juvenile more susceptible to being washed away. A more speculative scenario is that the horseshoe crab had been carelessly dropped into the lagoon by a predator, such as a pterosaur flying above; numerous pterosaur species have been found at Solnhofen. Tantalizing as this hypothesis might appear, it can be ruled out, as there are no predation marks on the horseshoe crab.

A ruffled surface, consisting of multiple foot, leg, and tail marks, along with circular depressions of the head shield, identifies the start of the track. This represents the point at which the horseshoe crab was flung into the lagoon, perhaps during a storm, fell through the water, landed on its back, and struggled to firmly right itself. Living horseshoe crabs swim upside down, a behavior often seen in juveniles, and tend to come to a rest on their back. Lying upside down, they rock their body side to side and use their tail to correctly orient themselves, sometimes leaving a circular depression in the sand (from the head shield), which would explain the ruffled surface present at the start of this fossilized track. Rotated and ready to march, the horseshoe crab made its way across the dark lagoon floor, where its final movements in time were recorded in the soft mud.

FIGURE 3.7. (Opposite). *The final descent.*

Having been flung into the lagoon and splashed through the water's surface, this little limulid is sinking rapidly into the deadly waters below.

BOB NICHOLLS 2020

The track is not a straight line. From beginning to end, it changes from straight to meandering, often with sudden movements, including total body turns, showing how the horseshoe crab's walking style altered. One particularly noticeable feature present throughout the track is the use of the long spiny tail, which is identified by a single straight line positioned between the footprint impressions on either side. This shows that, apart from when the horseshoe crab lifted its tail when it changed direction, the tail was in contact with the soft mud for the duration. However, toward the end of the track, the tail impressions become shorter and more sporadic, indicating that the juvenile was constantly stopping and raising its tail, presumably because it was distressed and disoriented.

In conjunction with this behavior, at certain points in the track, especially toward the end, the leg imprints are deeper and indicate where the horseshoe crab likely attempted to push off the soft mud in the hope of swimming to freedom but had little remaining energy to do so. The change in walking behavior shows how the horseshoe crab began to struggle in the oxygen-free toxic trap and slowly dragged its near-lifeless body through the mud before eventually suffocating and dying in its tracks.

Having the entire trackway preserved from beginning to end, along with the animal that made it, is an entirely unique association of the final end-of-life moments of an ancient animal. This specimen represents an incredible snapshot of one juvenile horseshoe crab's ability to momentarily tolerate the perilous nature of a deadly lagoon, only for its inexperience to ultimately lead to its demise.

MASS MOTH MIGRATION

Moths and butterflies are among the most beautiful and instantly recognizable insects. Together they form the large insect order Lepidoptera, which includes a staggering 180,000 living species, although new species are found every year, and estimates suggest that there might be as many as half a million.

As for fossil remains of these insects, it is a different story. They are particularly rare; only a few hundred prehistoric species have been described. The oldest fossils are from the Late Triassic and are around 205 million years old, so the paucity of specimens seems odd given the length of time they have been around. This is not an indication for low species diversity; rather, it is a reflection of their lifestyle and the generally poor preservation potential of lepidopterans as fossils.

Living species can form immense groups or swarms of thousands or even millions, and they migrate seasonally in search of favorable weather conditions, better food resources, breeding grounds, and much more. A group of moths is called an "eclipse," and a group of butterflies is referred to as a "kaleidoscope." Migratory species are present on all continents except Antarctica, and some migrate incredibly long distances, thousands of miles, crossing seas and continents. This makes you wonder whether prehistoric species migrated, too. Presumably they did. However, with such a poor fossil record, it would seem highly unlikely that an answer could be found. Or would it?

In Denmark, there is a series of rocks referred to as the Fur Formation, named after the small island of Fur, which are 55 million years old from the Eocene, approximately 10 million years after the extinction of the nonbird dinosaurs. A wide variety of exceptionally preserved animals and plants have been discovered in this formation, including something entirely unexpected: a collection of about 1,700 individual moths. When this find was first announced, in 2000, the discovery more than doubled the known number of moth fossils.

This large group included complete individuals, wingless bodies, and isolated wings from at least seven distinct species. The vast majority (of more than 1,000) belonged to a species with a body length of just 14 millimeters and were identified as members of a group of living lepidopterans, the Heteroneura. Significantly, individuals of this species were often found together in great abundance throughout several horizons in the Fur Formation. The exciting thing is that these moth-rich rocks originally accumulated over an offshore area of the ancient North Sea—evidence that these moths were at sea in mass numbers, away from their coastal habitat.

Several butterflies and moths migrate across the North Sea today, and there is a greater chance that more individuals are caught offshore when the winds are calm and temperatures on land are high. It is likely that the prehistoric moths embarked on their flights during similar conditions.

The huge number of specimens and their discovery in rocks of marine origin indicate that these prehistoric moths undertook mass migrations over the ancient North Sea. Further support for this interpretation was presented in 2017, after a similar discovery was made in slightly younger rocks of the Zagros Mountains in Iran, where fossil locusts were found in deep marine rocks, indicating that they, too, were part of a mass transoceanic migration. As the moths were found in large numbers at different levels in the rock formation, this suggests their flights were not a singular or local phenomenon, but instead occurred at different intervals in time. This unique discovery represents what must have been a common occurrence, just as occurs today.

FIGURE 3.8. (Opposite). *A whisper at twilight.*

An eclipse of moths journeys across the ancient North Sea as part of their annual migration. Some are taken off guard by a sudden gust of wind and fall into the watery depths below.

DINOSAUR DEATH PITS

If you want to impress somebody with a dinosaur fact, look no further than the sauropod dinosaurs. This group of long-necked, long-tailed behemoths grew to enormous sizes, upward of 30 meters and 70 tons; they weighed as much as seven fully grown African elephants. And, yes, they walked and lived on land. As the biggest dinosaurs ever, and the largest animals to walk on Earth, it can be expected that sauropods would leave the biggest tracks. And they did. Sauropod footprints are found globally, from extensive trackways exposed in the Jura Mountains on the French-Swiss border, to giant meter-long tracks in Western Australia. With so many discoveries it is understandable that something out of the ordinary might occasionally crop up.

Sauropods are not typically seen as killer dinosaurs but, rather, as plant-guzzling giants. These herbivores required huge amounts of foliage to sustain their (often) gigantic bodies, which acted as their main form of defense. Sauropods must have defended themselves from would-be attackers, and doing so probably led to the death of many predators. Maybe one day the fossil record will surprise us with such a fossilized victim, but for now, that is merely a figment of our imagination. However, an unusual discovery revealed in 2010 shows that some sauropods were, in fact, killers, albeit unintentionally.

During the Late Jurassic, about 160 million years ago, an enormous, 25-meter long, 20–30-ton sauropod called *Mamenchisaurus* trampled through deep, soft mud in an ancient marshland. A series of pits, between 1 and 2 meters wide and deep, were left behind as the sauropod pulled its sunken feet from the mud. These pits quickly filled with sediment and water and became unlikely death traps for smaller animals, who accidentally fell in and could not escape.

This interpretation is for three unusually preserved vertical bone bed pits found in the Wucaiwan area of the Junggar Basin in Xinjiang, China, the first of which was collected in 2001. They were found in rocks that show evidence of a warm, seasonally dry wetland environment that was frequently covered by volcanic ash. Analysis of the pits shows that they contain a soft sediment mixture of soil and volcanic mudstone and sandstone and at the time were filled with water-saturated mud.

Although a *Mamenchisaurus* was not found with its feet inside the mud, the consistent shape, size, and depth of the pits indicate that they were left by a large dinosaur, the largest of which to be found in the area and in the same rock formation include sauropods like *Mamenchisaurus*. Supporting this assertion, similarly preserved pits of varying sizes are common throughout the area, and several have been found as distinct linear trackways. However, of all the pits examined, only the aforementioned three have so far been found to contain fossils.

Once filled with watery mud, these deep pits would have appeared stable from the surface, but any small animals that unwarily stepped into them would have struggled to get out. We know this because buried inside these pits are the remains of numerous animals, including articulated and associated skeletons of at least eighteen small theropod dinosaurs.

The theropod skeletons belonged to three types, all new to science at the time, the most common being the short-armed herbivorous ceratosaur *Limusaurus*. (Not all theropods were carnivorous.) One of the pits contained at least nine *Limusaurus* of different ages, along with several skeletal remains of small mammals and reptiles, including a turtle and two small crocodilians. The largest dinosaur found buried inside one of the pits was *Guanlong* (represented by two skeletons), an early crested relative of *Tyrannosaurus rex*. Unlike its much later gigantic cousin, *Guanlong* stood just 66 centimeters tall at the hip and was unable to touch the firm base of the pit even with its feet, the pit depth being at least twice the height of this dinosaur.

Like layers in a cake, most of the skeletons preserved in the pits are vertically stacked on one another, separated by 5–20 centimeters of rock. This shows that the dinosaurs fell into the pits at different times and that those

FIGURE 3.9. (Overleaf). *In the wake of colossus.*

Several *Limusaurus* call for help as they are trapped and unable to escape from the deep and muddy tracks left behind by a giant *Mamenchisaurus*. Intrigued by their calls, a tiny tyrant, *Guanlong*, is attracted by the prospect of an apparently easy meal but will soon find itself in danger.

BOB NICHOLLS 2020

at the bottom were obviously buried first. Several of the dinosaurs are represented by partial skeletons and isolated body parts, which indicates that some of the carcasses must have been exposed and rotted for a short while before burial, floating partially submerged in the mud, probably for no more than days to months. Some evidence, in the form of broken bones in an otherwise perfect skeleton, suggests that as the bodies began to stack up, other animals that fell into the pits were able to get out by stepping on the dead bodies that lay beneath, hidden in the mud.

These dinosaur death pits capture an unexpected circumstance formed under unusual conditions. A sauropod taking a simple stroll through a muddy marsh resulted in the demise of multiple small theropods and other animals that became trapped in the mud-filled footsteps of giants.

TIME TO GROW WHEN YOU MOLT AND GO

Have you ever seen an arthropod shed its old, worn-out exoskeleton? Cracking out of and emerging from its lifeless former self is one of the most surreal acts of nature that all arthropods do, whether spiders, centipedes, locusts, or lobsters.

Called *ecdysis*, or molting, the shedding of the exoskeleton (the outer cuticle) is vital for arthropods to grow bigger and stronger. Even damaged or missing appendages can be entirely regenerated. Once the shed shell (exuvia) has been cast aside, the body is initially soft, but it hardens shortly thereafter. Arthropods are at their most vulnerable during and immediately after molting, which can be fatal, because either the animal got stuck or it was preyed on. Most arthropods continue to molt throughout their lives, although the majority of insects and arachnids stop molting when they become adults.

Arthropods are among the most ancient of all living animals and have an extremely rich fossil record, reaching back almost 540 million years. The inference has always been that early arthropods must have molted, especially as molting is ubiquitous among the more than one million (and counting) species alive today. But if this is the case, how could we determine whether said fossil arthropod was the actual deceased individual (a carcass) or an exuvia?

Understandably, this can be very difficult to decipher. Generally, if the fossil is complete with no signs of damage, there is a good possibility that it is the carcass and not the exuvia. Those that show disarticulation—especially between the head (cephalon) and body (thorax)—and display distinct breakage lines can indicate the location where the individual broke free from its old exoskeleton. In all likelihood, we can expect to find more fossil molts, because arthropods can produce numerous molts in their lifetime but leave behind only one carcass. Still, it is hard to distinguish a molt from a carcass because the fossilization process can often damage and distort specimens and therefore provide a false picture.

The best way to determine the difference is to find a specimen in the act of molting. This is more easily said than done. However, considering that molting might be responsible for up to 80 to 90 percent of living arthropod deaths, there appears to be a good chance that this behavior would be captured in the fossil record.

Remarkably, rare specimens have been found. The earliest unequivocal arthropod fossil caught in the act of molting is approximately 518 million years old. It was only recently discovered, in 2019, from the Early Cambrian Xiaoshiba biota of Kunming, southern China. This early marine arthropod is an example of a species called *Alacaris mirabilis* and is preserved mid-molt, with its partly discarded exoskeleton still attached, overlying the carcass of the newly emerging individual.

The first Cambrian critter to be found in the act of molting, however, was a 505-million-year-old specimen of *Marrella splendens*. This tiny, 2-centimeter-long fossil is from the famous Burgess Shale in British Columbia (see chapter 2 on arthropods and their eggs), where more than 25,000 *Marrella* specimens have been found. Unfortunately for this *Marrella*, it made it only halfway through the act of molting. As preserved, its antennae and part of the head are poking out from the old exoskeleton, emerging from the point where the head meets the body, with the remainder stuck inside. Intriguingly, its wide lateral head spines are folded inward and backward, the opposite of that found in normal *Marrella* specimens, which indicates that they were softer during ecdysis.

As amazing as these earliest known molting arthropods are, the most striking examples are from the Jurassic Solnhofen limestone in Germany, collected near the same area as the horseshoe crab death march. Although most animals preserved in the Solnhofen lagoons succumbed to the anoxic bottom waters, like the horseshoe crab, some of these lagoons (even at the bottom) were habitable and could support life, at least temporarily.

Captured on a slab of limestone is the entire process of a molting arthropod. Created by a lobster-like crustacean called *Mecochirus longimanatus*, the fossil begins with a distinct landing trace, identifying the moment the crustacean fell through the water column and came to rest on the lagoon floor. It then crawled about 30 centimeters, pushing itself along with its tail fan. Attempting to remove the exoskeleton, thrashing and twisting around, it created a series of bent furrows, ridges, and scratches in the sediment, some of which were made while it lay on its side. Finally, it emerged from its shell and walked away, leaving its neatly preserved exuvia behind.

It is not possible to say with certainty how long the molting process lasted, because the time varies in living arthropods, although it commonly takes several minutes. However, comparisons can be made with modern decapod crustaceans like lobsters and shrimps, who toss and turn to rid

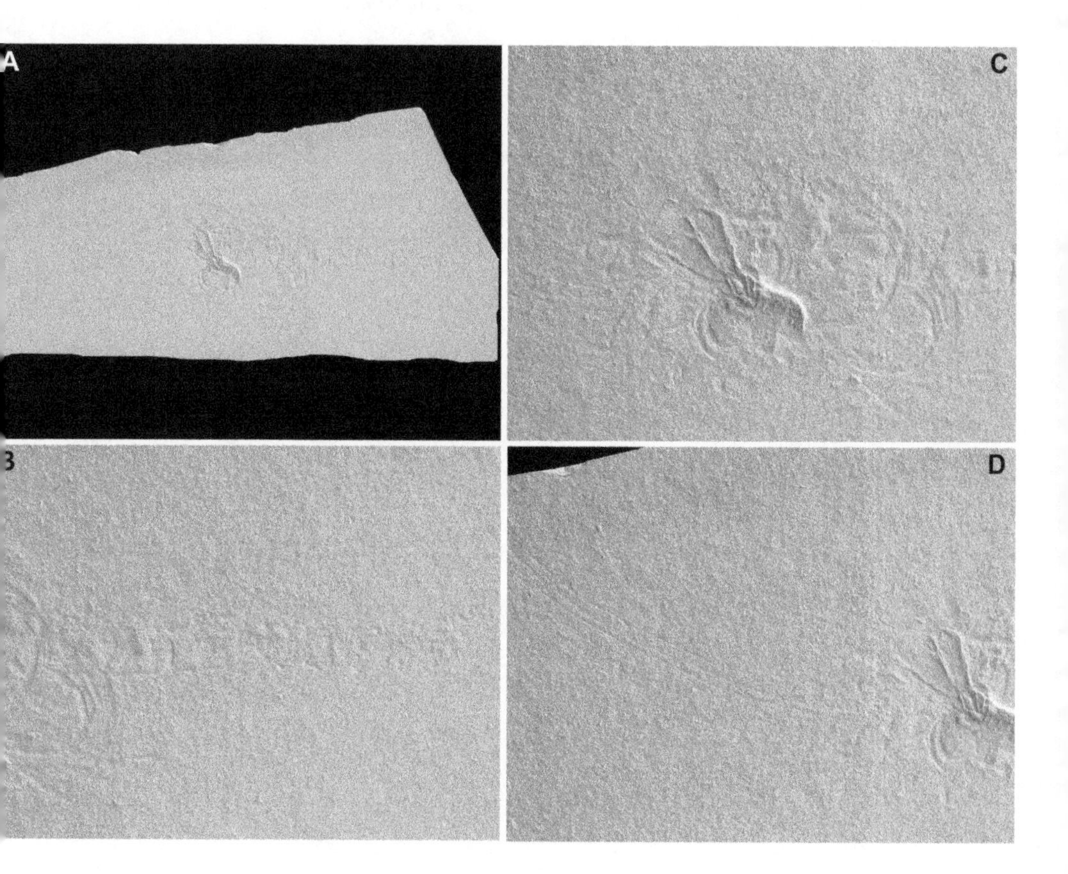

FIGURE 3.10. (*A*) Complete overview of the trace and exuvia left behind by a *Mecochirus longimanatus*. (*B*) Beginning of the trace where the crustacean landed and walked a short distance. (*C*) Various well-defined marks from tail and appendages, showing where *Mecochirus* was in the process of molting (with shed exuvia left behind). (*D*) Walking trace showing that the fresh-faced crustacean left the scene. From the Solnhofen lithographic limestones of Langenaltheim, Germany.

(Courtesy of Günter Schweigert and Lothar Vallon)

themselves of their old exoskeletons. In fact, the traces made by *Mecochirus* resemble the molting process of the living sand lobster, especially in how it rolls onto its side to aid with the removal. In this modern analogue, ecdysis can take around thirty minutes, after which it takes a further ten to twenty minutes for the sand lobster to gain mobility and move on. This provides a possible time frame for the molting movements left by this 150-million-year-old Jurassic relative.

The Cambrian arthropods caught in the act of molting confirm that ecdysis occurred early in arthropod evolution, something that, up until their discovery, had been only inferred. Capturing the complete molting sequence, the spectacular *Mecochirus* and its trace provides a fascinating glimpse of a crucial behavior in the life of this ancient arthropod.

FIGURE 3.11. (Opposite). *Molt and go.*

Mecochirus longimanatus finds a quiet place to eject its old, worn-out exoskeleton, leaving all the clues behind as it flees the scene to live another day.

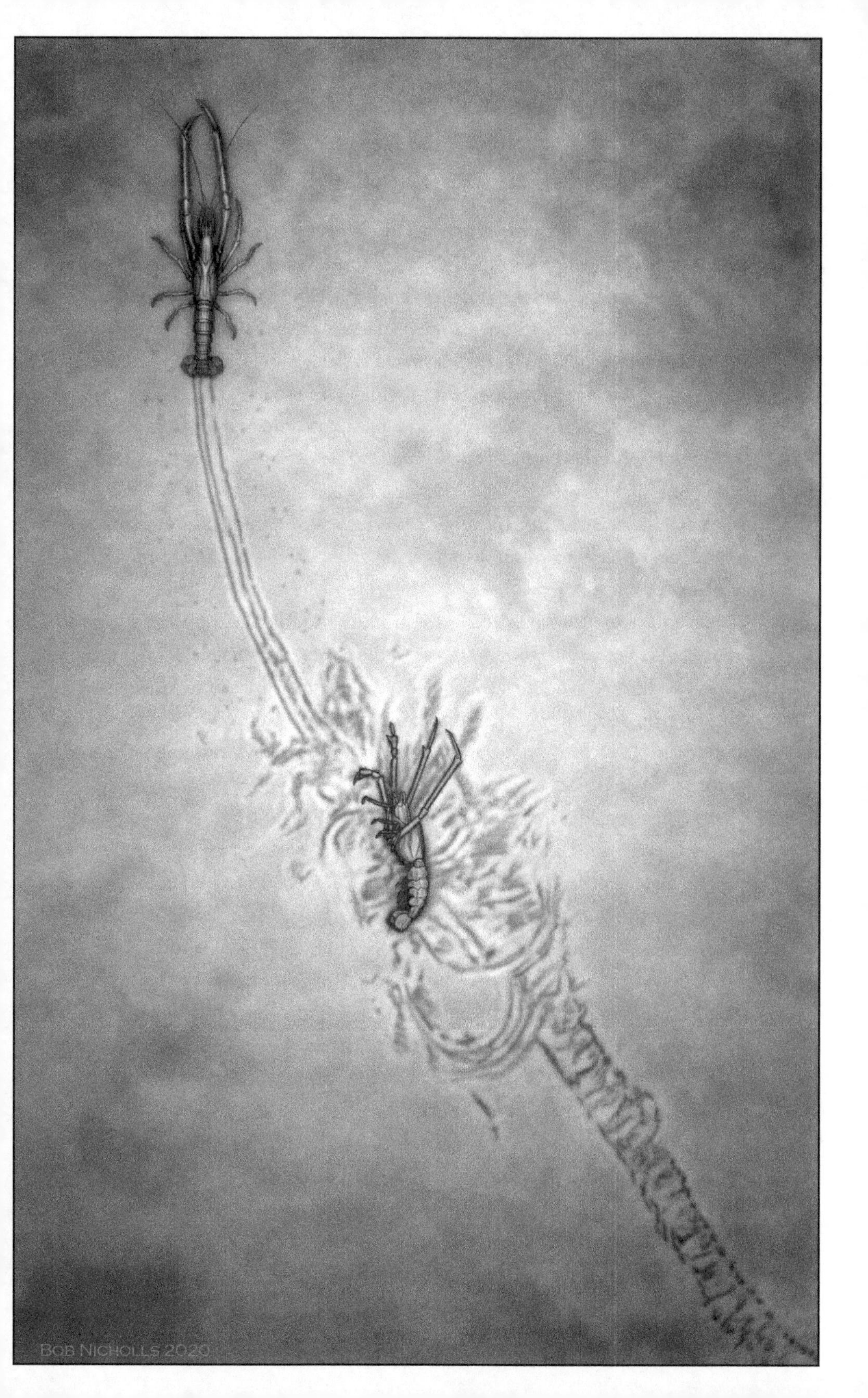
BOB NICHOLLS 2020

PREHISTORIC PARTNERS: THE ODD COUPLE

Numerous fossilized burrows from Early Triassic–aged rocks have been found in southern Africa, especially the Karoo region, a vast semi-desert that has yielded thousands of fossils. They are around 250 million years old, which is about the same time that the planet faced its greatest natural disaster. This was a major, global cataclysmic disaster, referred to as "The Great Dying," in which about 90 percent of all life was eradicated at the end of the Permian. What triggered this worldwide extinction is still debated, but major volcanic eruptions and an asteroid impact definitely played a critical role in Earth's most severe extinction event of all time.

A major change in the climate resulted in global scorching hot desert-like conditions. Adapting to the harsh environment, some of the land-living vertebrates made burrows to escape the extreme heat. Many of these Early Triassic burrows have been collected from the Karoo and are thought to represent a direct behavioral response to the changing climate. Burrows act as an excellent refuge, a great way to avoid predators, and can serve as protection from harmful weather conditions and enable animals to live in the comfort of their own home. Although plenty of these fossilized burrows have been found, they are often empty.

In exceptionally rare cases, articulated skeletons of early mammal relatives called *therapsids* have been found buried inside such burrows. A few of these contain individuals lying alongside each other, curled up and resting in the burrows they almost certainly created; sometimes scratch marks are also present in the burrows. It has been suggested that this burrow-and-occupant association is an indicator of seasonal dormancy. This is a neat theory because it suggests that these individuals were in a state of inactivity called *aestivation*, similar to hibernation, which is undertaken during a particularly hot and dry part of the year. That behavior would fit the extreme weather conditions at the time.

An unusual and unique animal association lay hidden inside one fossil burrow collected in the Karoo in 1975 from the Oliviershoek Pass in KwaZulu-Natal Province. When it was found, only a portion of a skull was exposed, which was tentatively identified as belonging to a relatively common, fox-sized therapsid called *Thrinaxodon liorhinus*. The specimen was broken into two parts, and more bones were found inside. As nothing out of the ordinary appeared to be preserved, it was placed in the fossil collections of Wits University in Johannesburg, and its secret was only recently revealed after reexamination of the fossil.

As this particular burrow cast remained unprepared, meaning the excess rock had yet to be removed from the fossil in the laboratory, and considering that it was packed with bones, the decision was made to examine the specimen without causing any damage or displacement. The burrow was taken overseas to France, where it was fired at with lasers. Specifically, it was scanned inside a synchrotron, a huge, powerful machine that produces incredibly detailed X-rays through the acceleration of electron particles moving almost at the speed of light. Using this nondestructive process meant that the bones would not be disturbed and a wealth of information would be revealed that could not be observed externally.

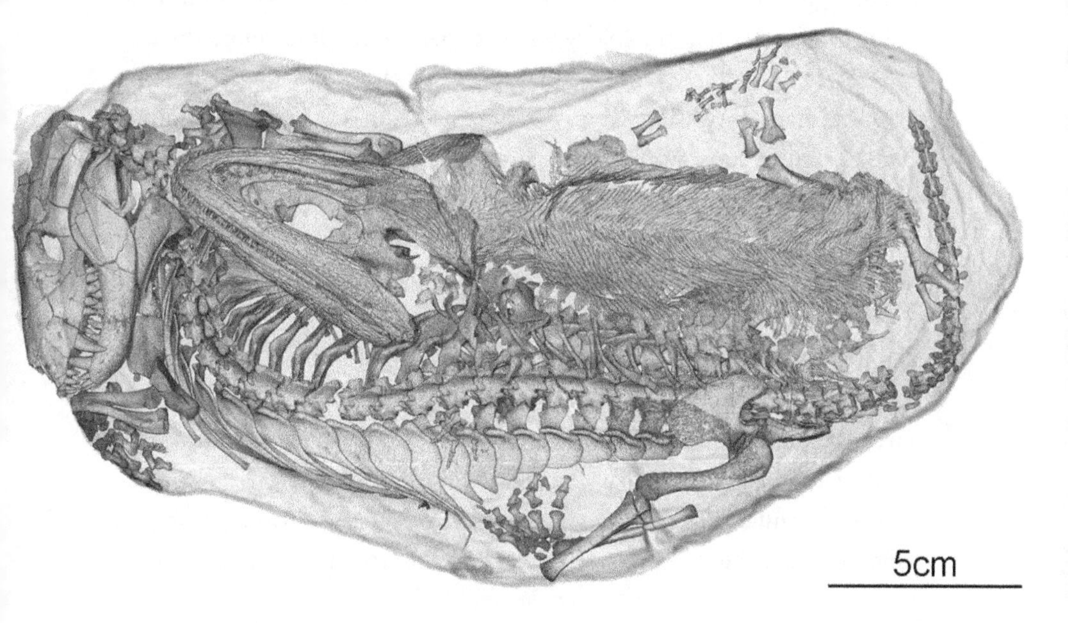

FIGURE 3.12. Time for sleep—a 3-D rendering of the fossilized burrow containing an injured *Broomistega* amphibian nestled up against an aestivating *Thrinaxodon*.

(Image from Fernandez, V., et al. 2013. "Synchrotron Reveals Early Triassic Odd Couple: Injured Amphibian and Aestivating Therapsid Share Burrow." *PLOS One* 8, e64978.)

Access to such equipment has helped to revolutionize how paleontologists study and analyze fossils. This type of technology can allow us to peer inside the rock and determine how complete or significant the fossil contained within might be.

The findings were remarkable. Encased inside this burrow, lying side by side the complete *Thrinaxodon* skeleton, was a complete and similarly-sized salamander-like, juvenile amphibian called *Broomistega putterilli*. Even the amphibian's patterned skin is still present. As preserved, *Thrinaxodon* is lying on its stomach with its head awkwardly twisted to the left, as if pushed against the end of the burrow, whereas *Broomistega* is lying on its back, exposing the underside, and resting against *Thrinaxodon*. This is a perplexing association of two very different animals, which raises intriguing questions regarding their interaction.

Interestingly, the amphibian bears a series of seven broken ribs on its right side. It could be assumed that the *Thrinaxodon* had attacked the amphibian and brought it into its lair. The broken ribs, however, show signs of healing and demonstrates that the injury was sustained some time before the pair met. In fact, it is likely that the injury was the result of a single crushing blow—perhaps it was stepped on—that happened weeks before the animal died. Certainly, such an injury would have affected the animal's ability to walk and would have caused severe pain, particularly while breathing. An injured amphibian struggling to walk in the baking heat of the sun was a meal waiting to be eaten.

Based on the anatomy of its skeleton, we know that *Broomistega* was adapted for a semi-aquatic lifestyle, but the structure of its limbs suggests it was unable to burrow. Conversely, *Thrinaxodon*'s limbs were adapted for burrowing, and, coupled with the discovery of other *Thrinaxodon* inside their burrows, suggests it had created the burrow. As there is no indication of a scuffle, *Thrinaxodon* was either sleeping (aestivating) or was happy to have the company and tolerated the presence of the intruder.

FIGURE 3.13. (Opposite). *Needs must.*

An injured *Broomistega* amphibian seeks refuge from the baking hot sun and finds a hole in the ground where a *Thrinaxodon* is comfortably at rest.

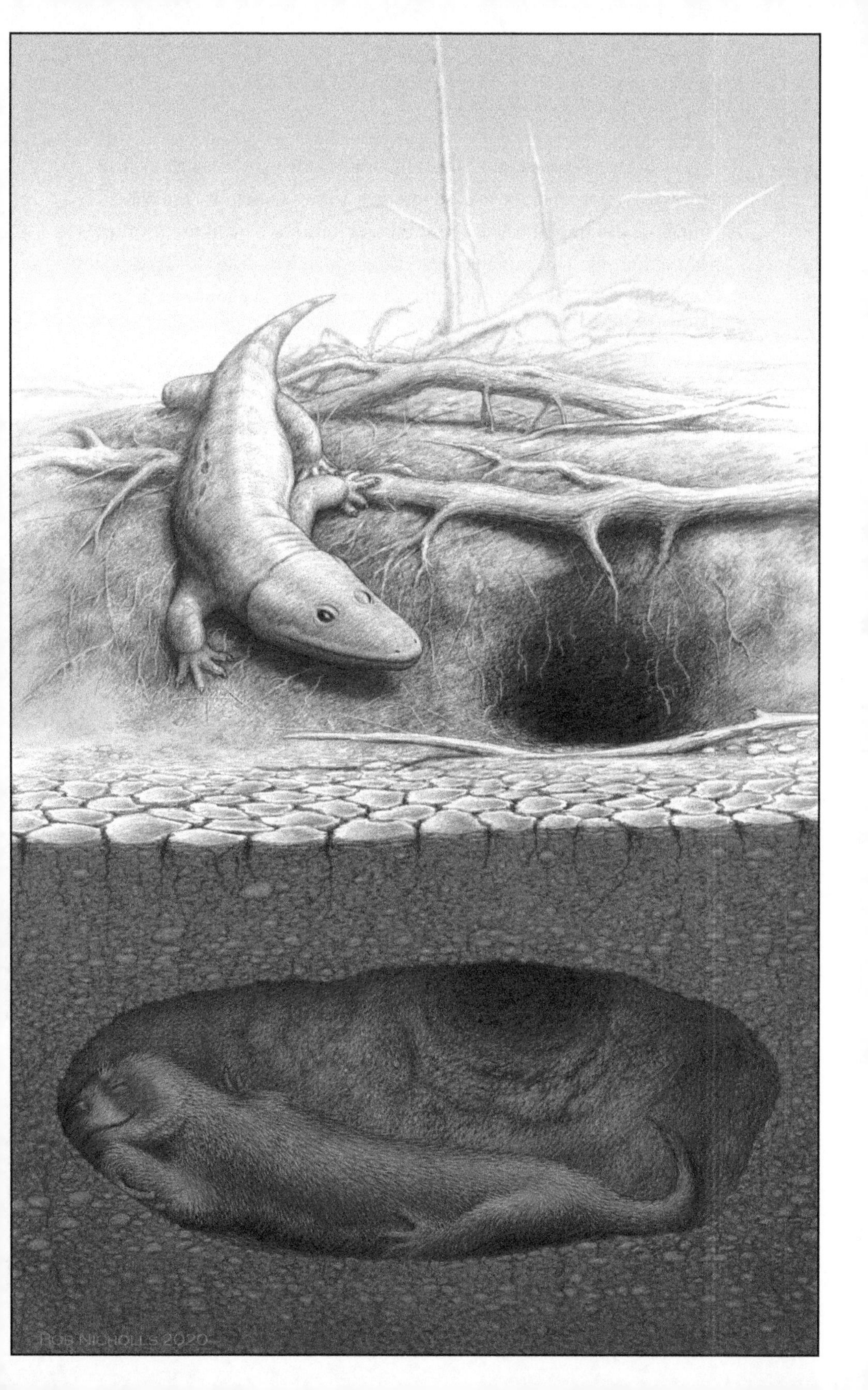
ROB NICHOLLS 2020

This relationship is an example of commensalism, which, as noted in chapter 2, occurs when distinct species interact, with one reaping the benefits (in this case, the amphibian) without causing harm or benefit to the other.

On the flip side, it is possible that *Thrinaxodon* was dead inside its burrow and that *Broomistega* simply moved in and later died, although their exceptional preservation and direct association suggests that they died together. Thus, following its instinct to survive, *Broomistega* sought protection and climbed into the burrow, where it also might have entered a deep sleep. Similar behavior has been observed in living amphibians, particularly juveniles, which sometimes seek refuge in other animals' burrows.

It is most likely that the *Thrinaxodon* was in a period of dormancy when the injured amphibian entered its burrow. Whatever the exact scenario that led to this association, the pair suffered the same ill-fated consequence, as the burrow rapidly filled with sediment during a flash flood, preserving them together forever. It provides a direct, unusual glimpse of communal interaction between entirely unrelated animals that simply would not have been considered were it not for this chance discovery.

DEVIL'S CORKSCREWS

Some fossils can be so peculiar and different from anything else that they can leave paleontologists scratching their heads for decades trying to figure out what they are. During the late 1800s, cowboys and ranchers across Sioux County in Nebraska began finding strange spiral sandstone structures protruding from the ground and cliffs. Many were taller than the average human, over 2 meters high, but nobody knew what to make of these puzzling structures. Locally they became known as "devil's corkscrews."

One by one, these sandstone helices continued to be found, and not until 1891 were they brought to the attention of scientists. Erwin Barbour, a geologist and professor at the University of Nebraska, was the first to examine and attempt to interpret these structures. He was quick to recognize that they were fossils and named them *Daemonelix* (or *Daimonelix*), the Latinized version of their nickname. Barbour clearly had a sense of humor, although his identification of these structures, which he believed to be the fossilized roots of a type of freshwater sponge or plant, caused quite a buzz.

Other scientists were not convinced of these identifications. Instead, they thought the large structures were fossilized burrows, because some were found to contain the fragmentary remains of rodents buried inside, which they assumed were the burrows' creators, an association that Barbour had dismissed. The discovery of additional, more complete finds, including the first identifiable remains of an extinct terrestrial beaver called *Palaeocastor*, found preserved in its burrow, proved this to be correct. These were the earliest mammal fossil burrows to be recognized and described.

The association between the burrow and its occupant (*Palaeocastor*) provided the first convincing piece of evidence that these prehistoric beavers, which were no larger than today's groundhog (woodchuck), were the creators of such absurd trace fossils. Since the original finds, numerous beavers have been discovered entombed in their helical homes. Yet, a mystery remained: how exactly did they create these unusual structures?

Almost a hundred years after the original discovery, an extensive study published in 1977, based on the examination of more than five hundred burrows, provided the answer. Aligning the walls of these burrows were numerous grooves that were found to match perfectly the size and shape of the beaver's large, ever-growing incisor teeth. Using their powerful incisors, they excavated the perfectly formed burrows, carefully drilling into the ground and creating these spiral staircases. The spiral acted as the long opening for the burrow and served as the sole entrance and exit for the beavers. At the base of the spiral, the beaver created a straight living chamber that was inclined upward (up to 30 degrees), away from the spiral; these chambers could be as much as 4.5 meters in length.

But why the strange shape? It is possible that these twisted burrows acted as additional protection against larger predators that otherwise might have been able to reach inside the burrow had there been a simple, straight opening. However, others have suggested that the shape could have been a response to the hot, dry climate of the time. The helical design would have been an ingenious air conditioning system, helping to control temperatures deep inside the burrow. Together these functions might have created the ideal environment for nesting and rearing young. Support for this behavior comes from the recent discovery of a giant spiral burrow in northwestern Australia created by the living yellow-spotted monitor lizard (*Varanus panoptes*), which uses such helical burrows exclusively for nesting.

Most of these burrows are around 20–23 million years old, from a Miocene-aged rock exposure called the Harrison Formation. Today, thousands of specimens are known from the badlands of western Nebraska and neighboring eastern Wyoming. Those that contain skeletal remains likely represent sleeping beavers whose burrows were filled with sand and silt, perhaps during floods from heavy rainfall, which carefully preserved the inhabitants. Even today, specimens continue to be found, and some are protected at sites like the Agate Fossil Beds National Monument, where several corkscrews remain in the hillside.

With the notable exception of a spiral, similarities are shared with the burrows created by modern-day prairie dogs, rabbit-sized rodents that are native to the Great Plains of North America, including Nebraska. These burrowers are members of the squirrel family and live in a complex maze of underground tunnels, often in large groups of hundreds or even thousands, called towns. Numerous chambers are found inside these protective homes and are used for different purposes, such as a place to store food, a nursery to rear young, or even designated toilets. By comparison, the spacing

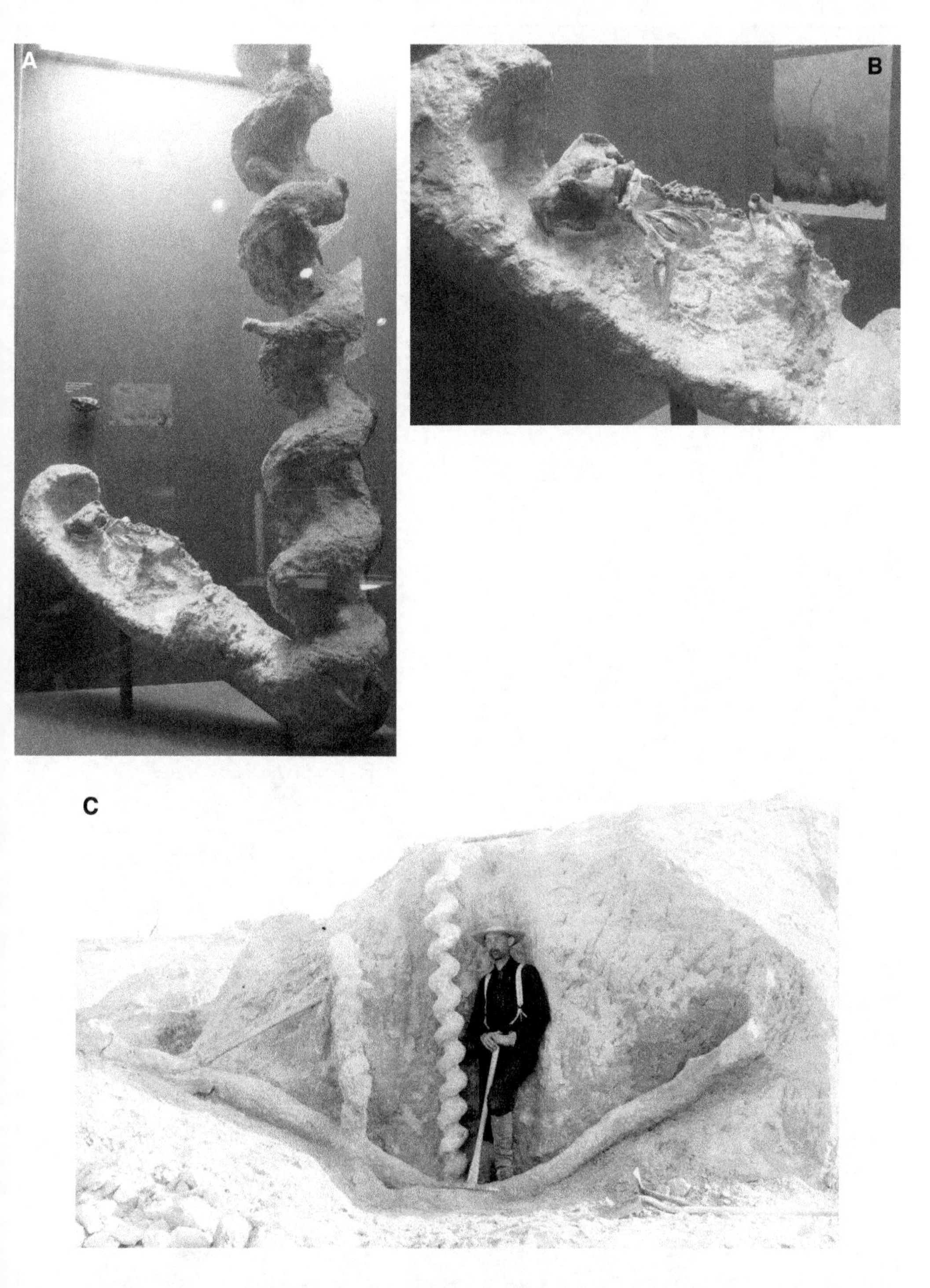

FIGURE 3.14. (*A*) The amazing feat of beaver engineering: the deep, spiral burrow home (*Daemonelix*) with its builder (*Palaeocastor*) buried inside. (*B*) Detail of the dedicated builder. (*C*) "Devil's corkscrews" discovered in the late nineteenth century at the Agate Fossil Beds National Monument, Nebraska.

([*A–B*] Courtesy of Wikimedia Commons; [*C*] courtesy of the University of Nebraska State Museum)

BOB NICHOLLS 2020

and number of many *Palaeocastor* burrows found in the same place shows that these prehistoric beavers had some form of community life and lived together, probably in such towns. Evidence from the remains of both adult and young beavers in the burrows suggests that they raised their litters inside the chambers of their spirally homes.

FIGURE 3.15. (Opposite). *The corkscrew family.*

A mother *Palaeocastor* cares for her young deep inside the chamber as the male keeps watch at the burrows' entrance.

THE DINOSAUR THAT LIVED IN A BURROW

Burrowing does not necessarily come to mind when we think of dinosaurs and their behaviors. It is difficult to picture a *Diplodocus* or *Tyrannosaurus* burrowing into the ground, however amusing that image might be. It might seem like an odd behavior for a dinosaur, but considering that species differ greatly in their size, shape, and habitat, among other things, the concept of burrowing dinosaurs is not so unsurprising. Since the 1990s, paleontologists have suspected that some dinosaurs were likely to have created and lived in burrows, although the direct evidence had been lacking.

It was in 2007 that a most wonderful discovery was revealed to the world. Collected from 95-million-year-old Cretaceous rocks near the Lima Peaks in southwest Montana was a large infilled burrow preserved with its occupants. Buried inside were the jumbled remains of three skeletons belonging to a new species of small-bodied, bipedal herbivorous dinosaur. Named *Oryctodromeus cubicularis*, meaning "digging runner of the lair," this animal had an anatomy that suggests it probably excavated the burrow using its short, powerful forearms and removed excess dirt with its snout. Here was a specimen made up of both the associated trace (burrow) and body (skeleton) fossils, for the first time providing compelling evidence of burrowing behavior in a dinosaur.

The sandstone burrow consists of a sloping, S-shaped tunnel that ends with an expanded terminal (living) chamber that contains the bones. Similar burrow systems are found in a wide variety of modern animals, such as the North American gopher tortoise, African aardwolf, and numerous rodents. Interestingly, the burrow has a tunnel width of 30–32 centimeters, a height of 30–38 centimeters, and a total length of just over 2 meters, although part of the burrow chamber was lost to erosion and so would have been longer. This indicates that it was a fairly tight fit for the largest *Oryctodromeus*, an adult with an estimated body length of 2.1 meters. Roughly two-thirds of the length, however, is made up of an extremely long tail, and there was enough space in the chamber for the individual to turn around. In living species, such tight-fitting burrows often deter predators, and some animals, like the aardwolf, create burrows whose tunnels are shorter than the animal itself so they fit snugly inside.

The two other *Oryctodromeus* skeletons are half the size of the adult and represent juveniles of the same species. It seems the adult ensured that the burrow was just large enough to accommodate it and the juveniles only, although traces inside the burrow made by much smaller animals (like insects and possibly tiny mammals) suggest that it might have shared the burrow. Given the size of the juveniles, this association suggests that *Oryctodromeus* cared for its young for an extended period inside the den. It therefore seems that parental care was, at least partly, a reason for the creation of the burrow, to ensure the survival of the offspring. Similarly, several species of living avian dinosaurs (birds) burrow underground and care for their young, such as the burrowing owl, found in the Americas, and the Atlantic puffin.

The geology of the rock formation where the burrow was found represents a prehistoric floodplain. Given the jumbled nature of the skeletons, some might argue that rather than representing a small family of dinosaurs in their burrow, the animals were washed inside by a river or flood or possibly might have been dragged into the lair by a predator. However, the completeness and association of the carcasses buried deep within a narrow burrow and the lack of bite marks or damage to the bones rule out such scenarios. Instead, *Oryctodromeus* burrowed into the mud and clay that later became filled with sand during a flood. The mass of mixed-up bones suggests that these individuals died (for reasons unknown) and decomposed in their burrow before they were buried.

Numerous *Oryctodromeus* skeletons have since been reported from eastern Idaho, where it represents the most common dinosaur from that state. Additional discoveries were also made in the Lima Peaks area. Several of these new specimens were found in association and include individuals of different sizes and ages, supporting evidence for the social behavior of this species. Remarkably, at least two burrows, one from Idaho and one from Montana, matching closely the original burrow system in shape and size, were also found to contain the bones of *Oryctodromeus*.

BOB NICHOLLS 2020

Oryctodromeus represents the first known burrowing dinosaur. This provides solid evidence not only that some dinosaurs dug underground burrows, which protected the occupants from predators and extreme weather, but also that they cared for their young for extended periods inside dens. Such burrow–body associations are among the most exquisite evidence of dinosaur behavior in the fossil record.

FIGURE 3.16. (Opposite). *The bunker family.*

An adult *Oryctodromeus cubicularis* rests with two juveniles inside the burrow it created. A small herd of *Oryctodromeus* can be seen aboveground.

GIANT UNDERGROUND SLOTHS

Super-slow movers and laid-back tree-hanging snoozers, sloths are an intriguing group of arboreal herbivores found only in the tropical rainforests of Central and South America. These small mammals are represented by six living species that rarely exceed 5 kilos in body weight and are divided into two families: the two-toed (*Choloepus*; two species) and three-toed sloths (*Bradypus*; four species); the difference is not actually in the toes, but the fingers. The slowest mammals on the planet, these small tree sloths are expert climbers but extremely poor walkers, the total opposite of some of their extinct prehistoric cousins, which were enormous, lived on the ground, and dug burrows.

Known as ground sloths, numerous species and several extinct families have been recognized from discoveries across the Americas. Some of the first finds were made by Charles Darwin during his famous voyage on the HMS *Beagle*. The last species became extinct only recently, within the last few thousand years, and evidence suggests that they were probably driven to extinction by humans. Compared with living tree sloths, some of these extinct forms, like the South American *Megatherium*, were as big as an elephant but were bear-shaped and shaggy-furred, weighed as much as four to six tons, and reached up to 6 meters in total length.

A curious bit of history surrounds one of the first ground sloth specimens ever brought to the attention of science, and the first in North America. A fragmentary skeleton that was found inside a cave in West Virginia was studied by none other than Thomas Jefferson, the third president of the Unites States. Initially, Jefferson thought the remains belonged to an enormous lion that might still be living, and he shared his findings with members of the American Philosophical Society in 1797. The species was named after Jefferson in 1825 as *Megalonyx jeffersonii*.

When these giant sloths were first recognized, some people assumed that they slept upside down and hung from trees, like living sloths. That is the equivalent of an elephant hanging upside down in a tree! Even Darwin mocked this interpretation. Instead, many of these animals possessed huge, powerful forelimbs and massive hands with strong, widely curved claws that were suitably adapted for digging—clearly not hanging.

Lending support for the idea of giant digging sloths, in the 1920s and 1930s enormous, perfectly formed fossil burrows (termed *paleoburrows*) were found in South America, from the same areas and of the same age as several ground sloths. The association seemed too good a match for it to be a coincidence, but they were not the only giant megafauna that could have created such burrows. Other animals, including giant, car-sized armadillos, were found, too.

Dating from a few million to roughly 10,000 years old, well over a thousand of these straight or slightly sinuous mega-burrows have since been discovered. They are primarily located in Brazil, especially Rio Grande do Sul, and Argentina, notably in the area around Buenos Aires. Many are preserved as open, cylindrical tunnels so large that a person can walk inside, whereas most are partly or completely filled with sediment. Although they differ in size, the largest burrows are an astonishing 2 meters tall by 4 meters wide and can extend up to 100 meters! These are the largest burrows ever discovered. Some of the tunnels even connect with others, forming complex systems.

The enormous size of the burrows, which could match only that of ground sloths, supports evidence that (at least) the larger burrows were created by these giants. Moreover, the walls and roofs of some of the burrows are commonly covered with claw marks left by their creators. These grooves match closely the claws of the large second and third finger digits of species such as *Scelidotherium* and *Glossotherium*, two medium- to large-bodied (1–1.5 tons and the latter up to 3 meters long) ground sloths whose remains have been found in the same area. Other larger and heavier ground sloths, such as *Lestodon*, might have constructed the biggest burrows, whereas most of the smaller burrows were likely to have been created by giant armadillos.

Some of the burrows bear distinctive smooth spots where the sloths rubbed their fur against the wall by brushing past the same spot on a regular basis or perhaps scratching the occasional itch. Modern large mammals, like elephants, tend to showcase such behavior, rubbing against trees or rocks to scratch an itch or rid themselves of parasites.

Given the large size and length of many burrows, it has been suggested that some were probably excavated by multiple sloths, perhaps over several generations. Very rarely have such burrows been found to be associated with ground sloth body fossils, but at least one from Argentina contained a juvenile and adult specimen of *Scelidotherium*.

In addition to the burrows, a wealth of fossil remains belonging to ground sloths have been found in natural cave systems in South and North America. Besides their skeletons, incredibly, mummified ground sloth skin with yellowish- or reddish-brown fur has been discovered. Some even show that they had small bony ossicles embedded in the skin, which probably acted as a form of armor-like protection against predators. Their fossil dung has also been found. When it was first examined it was so fresh—both to

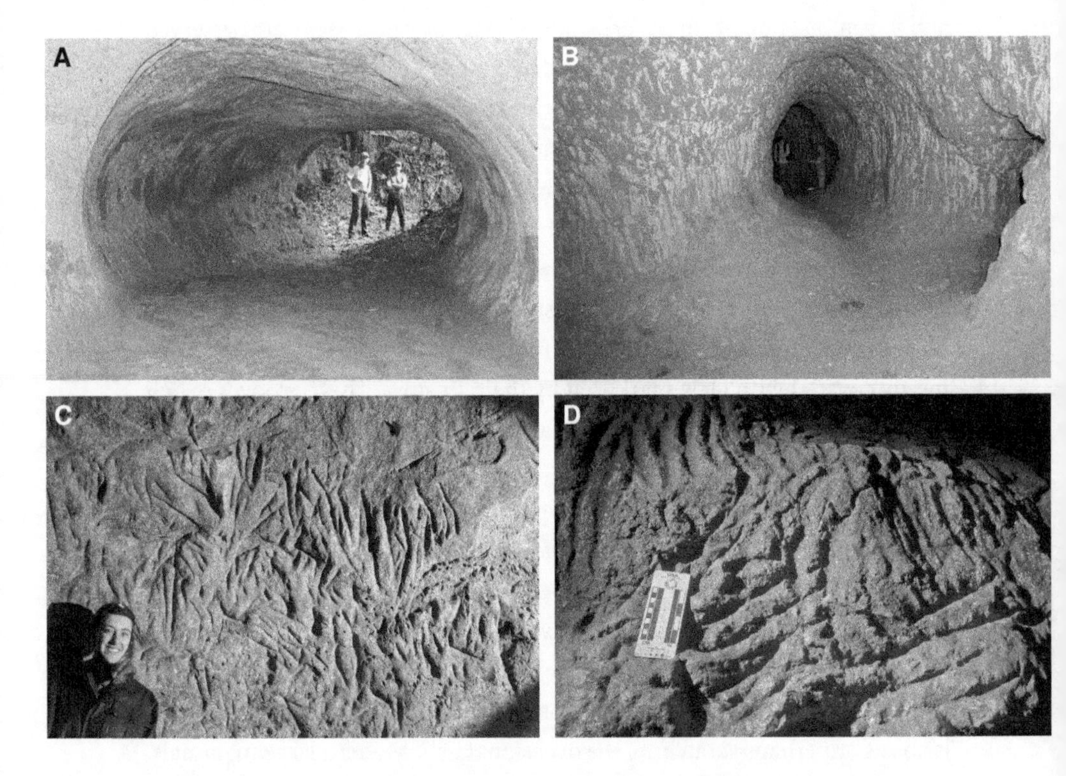

FIGURE 3.17. (*A–B*) Examples of enormous underground tunnels (paleoburrows) created by giant underground sloths, found in southern Brazil. (A is located in Timbé do Sul, in the state of Santa Catarina, and B is located in the state of Rondônia). (*C*) Various scratch marks in the wall of one of the tunnels matching the claws of ground sloths; from the same tunnel as in A. (*D*) Close-up of some scratch marks from a tunnel in Urubici, Santa Catarina state. Humans for scale.

([*A, C–D*] Courtesy of Heinrich Frank; [*B*] courtesy of Amilcar Adamy, Geological Survey of Brazil)

look at and smell—that its discoverers thought it was made by a living animal. Analysis of such dung has revealed the diet of several species.

You might wonder whether these giant ground sloths were also extremely slow movers. If they were, it would have taken an eternity to excavate and create such extensive burrow systems. That would be absurd. Plus, in living tree sloths, such super-slow speed is a form of *protection*. Their primary predators, like jaguars and harpy eagles, rely heavily on sight and movement to detect prey. Moving so slowly as to go unnoticed, sloths can often sneak by, camouflaged in the environment. There was no blending in for the ground sloths. In many cases, their immense size and large claws would have been their form of protection. Their digging abilities and even preserved footprints that show they moved faster than their living relatives indicate that these ancient sloths were not seemingly motionless giant chunks of vulnerable meat.

Why did these ground sloths go to all the effort of digging such enormous burrows? For one, to avoid predators. Sure, some species were large and able to protect themselves, but we know from evidence that humans hunted them, and juveniles would have been fairly easy targets for fleet-footed predators like saber-toothed cats. So, these burrows would have been a safe place for the sloths, perhaps even enabling them to live in families or communities. It also seems likely that they used the burrows to shelter from unfavorable weather conditions, be it colder or dryer environments.

Although these mighty movers are sadly no longer with us, their legacy lives on in a surprising place, the avocado. You see, ground sloths were around at the same time that avocado trees appeared, and as such they were among the largest animals that were capable of gulping down the fruit whole, pooping out their seeds, and helping to distribute the trees far and wide. By the time the sloths and other large avocado-eating mammals became extinct, they had already unknowingly helped to secure the future of the fruit. The next time you eat an avocado, you can thank these giant furry mammals.

FIGURE 3.18. (Overleaf). *Natural born diggers.*

Below an avocado tree, two young *Lestodon* are playing in the dirt that was excavated from the nearby burrow; one is practicing digging. The *Lestodon* parents squeeze out of the mouth of the burrow, where the one on the left is scratching its back. In the background, another adult is in the process of creating a new burrow.

Bob Nicholls 2020

4 Fighting, Biting, and Feeding

A great white shark darts through the water as if locked onto and pursuing a target. The hunt is on, though something is amiss. There is no sign of a seal, fish, or any would-be prey. This apex predator is not the hunter but the hunted. Cruising hot on its fins is a small pod of orca. Chasing down their prey and working cooperatively, these killer whales trap the great white and kill it. Instead of tearing the shark apart, they are after one thing—the large, calorie-rich liver—which they extract and consume, and then move on.

Without doubt, the great white and the orca whale are among the world's largest living apex predators. Whenever they meet, there always appears to be one clear winner: the orca. The idea that the mighty great white has a natural predator seems absurd and contradicts the image that is so deeply ingrained in our minds of this hunter sitting at the top of the food chain. Orca in the act of hunting and killing great whites has been caught on camera, and research has shown that the sharks will change their feeding habits and flee their favored hunting grounds when orcas are in the vicinity.

Watch any wildlife documentary, and the big sequence usually surrounds an intense fight, suspenseful hunt, or other type of classic predator-versus-prey situation, whether that involves the tiniest insects or the largest mammals. Why, you ask? Because on their own, behaviors like mating, nest building, or caring for young do not guarantee gripping TV drama the way that teeth chomping, claws slashing, and horns stabbing can. Watching a pack of predators dismember their prey, seeing an animal stand victorious over its competitor following a bloody battle, or learning how a certain animal has evolved an ingenious technique to catch and consume food grabs our attention and fuels our imagination. Of course, we know the animal world is considerably more complex and that nature is not always so "red in tooth and claw," as Tennyson put it.

It is a fact of life that every animal dies, and many fall victim to predators. During the brief moment it takes you to read this sentence, countless predators the world over will have made a kill. However, when there is a chance that you might be beaten or eaten, you have a choice: stand your ground and prepare to fight, or get ready to run like hell. This "fight or flight" response is driven by the animal's instinct to survive the threat of (possible) impending injury or death. If the animal survives—say, by startling and scaring off the attacker and maybe even killing it (think of situations like a zebra landing a lucky kick to the skull of an attacking lion), or by outrunning and escaping its foe—then it will live to see another day. Sure, luck can and does play a huge role in such situations, but the concept of survival of the fittest holds true for animals with the slightest advantage. Those strong, fast, and smart enough to overcome adverse scenarios are more likely to go on to find mates, have sex, and pass on their genes to the next generation.

Fighting and biting does not always mean feeding. Beyond predation, there are lots of reasons that animals fight, be it for dominance, learning, sexual partners, territory, shelter, or food. Disputes over such resources are common among conspecifics (members of the same species) and might end in death.

Take, for instance, the hefty African hippopotamus, one of the largest living land mammals. If you ever see one yawning, know that it is a warning. This is not a sign that they need a nap, it is a signal that you are too close. In the water, rival bulls might engage in savage fights over territory or females, clashing their gaped jaws together, pushing with power, and inflicting bloody wounds with their enlarged canine and incisor teeth. The fights might go on for hours and can have deadly consequences. As a side note, hippos are often thought of as strict herbivores, which is not true. Although it is rare, they have been observed eating animals and even resorting to cannibalism.

Biting is not always caused by fighting. Animals might use their teeth or equivalent parts for a variety of reasons, perhaps no stranger than during sex. You may well have witnessed a male house cat bite the neck of a female while mating, which he does to secure her and balance himself. Big cats like lions and tigers do this, too. In some species, sex, biting, and feeding typically co-occur. For instance, orb-web spiders mate, and then the female eats the male, but only if he has not yet escaped by cutting off his "penis"! Talk about a fatal attraction.

In paleontology, there is probably nothing more iconic than the image of a big toothy theropod tearing flesh from its kill. For many, the "fighting and feeding" angle is what really ignites their interest in animals from an ancient, bygone era. These types of questions pose a real challenge for paleontologists: "How did that dinosaur hunt?" "What were those long claws used for?" "Who ate whom?".

To help answer such questions, paleontologists utilize many scientific practices to draw hypotheses and often rely heavily on comparisons with living analogues. By observing who fights with whom, what eats what, and how these interactions occur in today's animal kingdom, we can learn a great deal about dominance hierarchy, predator–prey relationships, and food chains, along with various other roles that animals play in their communities and ecosystems. Naturally, understanding such interactions in prehistoric times can be far more complicated, and it might appear that only assumptions can be made.

Take *Tyrannosaurus rex*, for example. Specifics such as its immense size and bone-shattering bite, along with the fact that it is the largest carnivore found in its environment, would strongly suggest that this predator was at the top of the food chain, comparable to apex predators today. Similarly, by looking at how food chains work in living species, we can infer that a prehistoric butterfly was likely to have been eaten by a contemporary frog that was, in turn, eaten by an ancient bird.

Discovering a fossil with its last meal intact is not in the realms of fantasy. If the animal died shortly after it consumed its final feast, then the hard parts, like plants, bones, or teeth, should be present, but only if stomach acids had not eaten them away. Consequently, food remains are subjected to the same fossilization processes to which the animal that ate them was subjected and might also be preserved.

Though it might appear unlikely, a wealth of behavioral interactions concerning fighting, biting, and feeding are recorded in the fossil record. Just like detectives at a crime scene trying to determine "whodunit," paleontologists gather all the available clues and come to a justifiable conclusion. However, when we have direct evidence of animals engaged in fighting, caught biting, and found feeding on one another, these conclusions become undeniable. This chapter is not about the blood-thirsty, romanticized world of prehistoric life that some would have you believe. Rather, it is based on real evidence of dramatic moments contained in some of the most spectacular fossils.

CLASH OF THE MAMMOTHS

Watching two bull African elephants face off on the vast savannah is a sight to behold. As the world's largest land animals, weighing more than five tons, they engage in one of the mightiest, epic battles in the animal kingdom. This fight for supremacy often occurs when bulls enter an annual state known as *musth*, a time when their urge to mate is in overdrive, similar to deer in the rutting season. Temporarily pumped full of testosterone, as much as sixty times the normal levels, they become extremely aggressive and challenge other bulls for dominance, territory, and the right to mate with females.

Smashing their heads together, interlocking their tusks, and pushing with all their might ends with only the strongest elephant standing victorious. Sometimes these fights are to the death. All species of living elephant express similar displays of dominance, and this might suggest that their prehistoric ancestors did the same thing, or something similar.

In summer 1962, while walking in the grassy badlands of Nebraska near the small city of Crawford, a pair of workmen were surveying the area for the future construction of a dam. Little did they know that this routine trip would lead to a major mammoth discovery. Scanning the area, they stumbled on a large femur bone poking out from the side of a gully and sought an expert opinion.

It turned out that they had found part of the leg of a Columbian mammoth (*Mammuthus columbi*), one of the largest mammoth species, standing up to 4 meters at the shoulders. This was clearly an exciting find. Wasting no time, paleontologist Mike Voorhies, who was just completing his undergraduate degree at the University of Nebraska–Lincoln, was given the task of investigating what could turn out to be a complete skeleton. Voorhies asked around campus whether anybody wanted to come dig up a mammoth and assembled a team of young people, including college students and high school seniors.

Starting at 3:00 a.m. every morning and ending when the heat of the sun became unbearable, the team took just over a month to complete the dig. In the first few days, as the crew exposed more bones, it became clear that the skeleton was going to be complete. They began to chip away at the sediment surrounding the all-important skull and were initially disappointed with what they revealed. One of the tusks was facing the wrong way. The team contemplated whether the mammoth had somehow faceplanted and broken its tusk in doing so, which was frustrating, because they had hoped it would be a perfectly preserved skeleton. They continued to expose the skull, and only then did realization hit. That reversed tusk did not belong with the skull at all but to another mammoth.

Now, it is not uncommon to find mammoth remains associated together. In fact, several mammoth sites around the globe have revealed multiple mammoths, often preserved in large bone beds. But this site offered something unique: two mammoth skeletons preserved with their tusks tangled together. Had the team unearthed two bull mammoths fighting to the death?

Based on the roughly equal size of the skeletons, along with an assessment of their teeth (including the tusks), which are used to tell the age of an individual (just as in living elephants), they were identified as adults around forty years old. Although young male elephants will often play fight, they usually enter musth in their mid- to late twenties. Is it therefore possible that these two bulls were in a state of musth when they were fighting? Evidence from frozen mammoths found in Siberia shows they had specialized temporal glands, the same glands that are found on each side of the head in living elephants, which secrete a chemical substance when the bull is in full musth. So, it seems very likely that these bulls were locked in a musth battle, perhaps fighting for the affection of females.

When fighting, living elephants with straight tusks are able to stab their opponents and inflict much deeper wounds (using the tusk sort of like a spear), but those with curved tusks use them for pushing and interlocking and rely more on head-butting to do the damage. These mammoths' tusks were long and curved, making them unsuitable for stabbing; the tusks were primarily used to interlock, providing the necessary leverage to twist and push during combat.

The pair are preserved in direct contact, with their tusks twisted around each other. One of them has a complete right tusk but a snapped left tusk, whereas the other has a complete left tusk but a broken right. The stub tusks have blunt, rounded edges, which indicates that they were broken long before the fight erupted. The unusual tusk damage meant that the two were able to move in closer, without their tusks immediately clashing, and this is why they became stuck. Although rather gruesome, the tip of one of the tusks pokes into the eye socket of its opponent!

Locked together, both physically exhausted from all the pushing and pulling, after a final twist to shake loose, one of the mammoths slipped and dragged the other to the ground, and both fell to their eventual death. They fell in such an awkward way that their tusks overlapped and their bodies were trapped by each other's immense weight. (Each is estimated to have weighed ten tons or more.) It is reasonable to assume that one of the combatants might have been killed during the fight but was stuck with the other as a result.

It is unusual for living elephants to become tangled by their tusks, which are known to occasionally snap during fights. However, similarly, when elk and other deer fight during the rutting season, they can sometimes get stuck. In very rare cases, when the loser has succumbed due to the fight but is still attached, the winner will shake free its dead competitor and rip off its head in the process, inadvertently wearing the severed head as a trophy. If one of the mammoths did die, perhaps the victor was too tired and the loser too heavy that they could not separate. Trapped like this, they would have attracted predators, although none of the bones shows signs of scavenging marks.

Another unusual find was made only after the mammoths had been fully exposed and their bones carefully wrapped in plaster and removed for export to the Trailside Museum (where they are on display today). Found pinned underneath the front leg of one of the mammoths was a fossil coyote with a crushed skull. What was it doing there? Was the coyote caught in the most deadly of situations, between two battling bull mammoths, only to be trapped during the fight as the mammoth collapsed on top of it? Or was the coyote a scavenger, preying on the possibly dead mammoth when, suddenly, the living individual fell on top of it? Whatever the case, it seems likely that the coyote witnessed this clash of the titans only to become part of it.

Perhaps the most astonishing thing about this discovery is the extreme unlikeliness of its happening. The pair were obviously engaged in a mighty duel, became stuck together through strange circumstances, and were buried quickly enough—perhaps over years—that their skeletons remained intact, only to be found by complete chance. To this day, this fossil remains one of the most dramatic ever found, representing the largest prehistoric animals locked in their final battle to the death, facing off just as they were roughly 12,000 years ago.

FIGURE 4.1. The fighting mammoths, *Mammuthus columbi*, still locked in combat just as they were when they died. Each has one complete and one broken, much shorter tusk; the shortened left tusk can be seen clearly on the mammoth to the right. Note that the tip of the complete left tusk of the mammoth on the left is poking in the other mammoth's right eye.

(Courtesy of the University of Nebraska State Museum)

FIGURE 4.2. (Overleaf). *A fight to the end.*

Two Columbian mammoths in the height of musth are locked in combat, but locked together in such a way that they cannot separate. A single coyote has front-row seats for the mighty clash.

BOB NICHOLLS 2020

THE FIGHTING DINOSAURS

One of the most common questions all paleontologists are asked is which dinosaur would win in a fight between X and Y? As you would do when selecting your favorite character in a computer game, you have to rank each dinosaur. Who had the best weapons, the biggest teeth, largest claws, longest tail, and so on? It's a difficult question to answer, especially when the paleontologist in you wants to point out that often the two dinosaurs would have lived millions of years apart and never have actually met, so a lot of this is hypothetical and speculative. Nevertheless, this question captures so much interest because there is nothing alive today quite as immense or fierce as many of the dinosaurs.

Consider *Tyrannosaurus rex*, with its banana-sized teeth, and *Triceratops*, with its meter-long horns. It's easy to visualize what must have been one of the most incredible fight scenes in the animal kingdom. However, this clash of the giants is not a fantasy. It did happen around 66 million years ago, when both species lived alongside each other. Unfortunately, although we have some tantalizing remains of *Triceratops* with bite marks that match the teeth of *T. rex*, we have never found a *T. rex* and *Triceratops* fighting. (Although a yet to be formally studied fossil found in Montana might suggest otherwise.)

Of course, dinosaur duels would have been commonplace, just as we see living animals battling it out today. But knowing with absolute confidence that one species preyed on another is based largely on scant evidence or simply assumption. Sometimes, though, the most unimaginable fossil can be found. This was the case during an expedition in 1971, deep in the heart of the Gobi Desert in southern Mongolia. A Polish-Mongolian team of paleontologists collected several fossils, including one of the greatest and most famous dinosaur discoveries of all time—a pair of fighting dinosaurs.

Considering how unlikely it is for any animal to become a fossil, finding two largely complete dinosaurs literally fighting to their deaths is one of the most exquisite and improbable finds in paleontology history. It is also perhaps the most famous fossil that captures behavior in action. One part of this deadly pair is *Protoceratops andrewsi*, a boar-sized herbivorous cousin of *Triceratops*. Unlike its cousin, in addition to its small size, *Protoceratops* had a relatively small head crest and lacked the big brow horns so iconic of *Triceratops*. The other part of this pair is the predatory *Velociraptor mongoliensis*. *Velociraptor* is a name that needs no introduction, although it does require some explanation. Unlike in *Jurassic Park*, *Velociraptor* was about as tall as a turkey and probably weighed three or four times less than the much bulkier *Protoceratops*.

Because these two animals are locked in combat and facing each other, preserved as they were 75 million years ago, the evidence is overwhelming that they were captured in a battle to the end.

As positioned, the *Protoceratops* is crouched down with its body and head facing to the right, whereas the *Velociraptor* is laid on its right side with its head pointing forward. The left hand of *Velociraptor*, with its three curved claws, rests across the face of *Protoceratops*, perhaps scratching it, but the right forearm—just below the elbow—is clamped in the strong beak of the *Protoceratops*. If we add missing bones, flesh, and muscle, it's plausible that part of the right leg of the *Velociraptor* might have been trapped and crushed under the body of the *Protoceratops*. Incredibly, though, showing how this animal used its claws in close combat, the *Velociraptor*'s left foot is held high in the air, and its deadly, sickle-shaped claw is positioned deep within the throat region, where it might have delivered a fatal slashing blow to the *Protoceratops*. It would appear that the *Velociraptor* had the upper hand, but with its right arm trapped and right leg possibly stuck, escape was simply impossible.

Both these dinosaurs were exhausted and severely wounded. They lived in a desert with conditions similar to those of the Gobi today, so the consensus is that, perhaps due to heavy rains during a thunderstorm, a sand dune collapsed above the pair, flowing over them and burying them midfight in what was possibly a split second. Although this scenario seems most likely, it has also been suggested that the two were buried by a severe sandstorm or died as a direct result of the fight and were then slowly covered by drifting sand. Regardless, one thing that does not quite add up is that the *Protoceratops* is missing both arms, its left leg, and the end of its tail, yet the *Velociraptor* skeleton is complete. One suggestion is that the *Velociraptor* attacked and killed the *Protoceratops* but in doing so was trapped, eventually dying before being buried. Only later, predatory dinosaurs, perhaps other *Velociraptor*, found parts of the *Protoceratops* exposed and scavenged what they could. Additional evidence for such a feeding interaction, as the result of either scavenging or a group kill, has also been found in another example of *Protoceratops*, which displays various tooth-marked bones that were found in association with shed teeth matching those of *Velociraptor*.

Whatever the reason for their preservation, this spectacular discovery, considered a national treasure of Mongolia, is the first fossil to capture dinosaurs fighting to their deaths. Predator–prey interactions like this are simply astounding. Here, the final moments of two 75-million-year-old fighting dinosaurs are captured in time, providing irrefutable evidence that *Velociraptor* and *Protoceratops* battled to the death.

FIGURE 4.3. The famous, spectacularly preserved "fighting dinosaurs" fossil of a *Velociraptor* and *Protoceratops* in their fight to the death. Note that the left leg of *Velociraptor* is raised high and the infamous "killing claw" is positioned in the neck region of the *Protoceratops*.

(Reproduced with permission from Barsbold, R. 2016. "The Fighting Dinosaurs: The Position of Their Bodies Before and After Death." *Palaeontological Journal* 50: 1412–17, courtesy of Pleiades Publishing, Ltd.)

FIGURE 4.4. (Opposite). *The eternal stalemate.*

In an epic battle, a *Velociraptor* and *Protoceratops* fight to the death. Nobody wins.

JURASSIC DRAMA: A HUNT GONE WRONG

Soaring high above the warm, tropical sea some 150 million years ago, a small, toothy pterosaur was on the hunt. Called *Rhamphorhynchus muensteri*, this predator had spied a school of fish in one of the many shallow lagoons and went in for the kill. Diving into the water, it snatched a fish and swallowed it head first. Just at that crucial moment, as the fish slid down its throat, out from the depths came a swift predatory fish that attacked the pterosaur. The two became tangled before sinking to the bottom of the anoxic lagoon, where they were buried together forever.

Although this interpretation is hypothetical, it is one plausible scenario for a remarkable fossil. A complete pterosaur preserved with a tiny fish in its throat and a stomach full of half-digested fish remains is snagged in the spear-like upper jaw of one of the angriest-looking fish you will ever see, an 80-centimeter-long *Aspidorhynchus acutirostris*. With an elongated bill-like snout, sharp teeth, and long, slender body, this species looked a bit like a modern garfish. The fossil was found in 2009, collected from the Jurassic Solnhofen limestones near the town of Eichstätt in Bavaria, Germany. I have had the privilege of examining this specimen first hand, even visiting the quarry that it came from (I saw the hole in the ground) a couple of weeks after it was excavated. I can confidently say that it is one of the most exquisite fossils I have ever examined.

When the pair became intertwined, *Aspidorhynchus*'s teeth and pointed bill became jammed in, and had even punctured, part of the pterosaur's leathery left wing, which was formed by a membrane of skin that stretched from the tip of the wing to the ankle. In an attempt to break free, the fish would have vigorously moved its head, shaking the pterosaur from side to side, as is evidenced by the clear distortion of the pterosaur's left wing finger. (The rest of the *Rhamphorhynchus* skeleton remains intact and is naturally articulated.) Still tangled and fighting to break free, the fish might have swum deep into the lagoon and accidentally entered the poisonous anoxic waters, where it then suffocated. Presumably, the pterosaur had already drowned at this point. Of course, this is pure speculation, but whatever the case, they had to have drifted deep into the abyss within a short amount of time in order to be preserved, because otherwise they might have been eaten by something larger.

The weird association makes you wonder whether *Aspidorhynchus* was in hot pursuit of the same prey as *Rhamphorhynchus* and accidentally caught

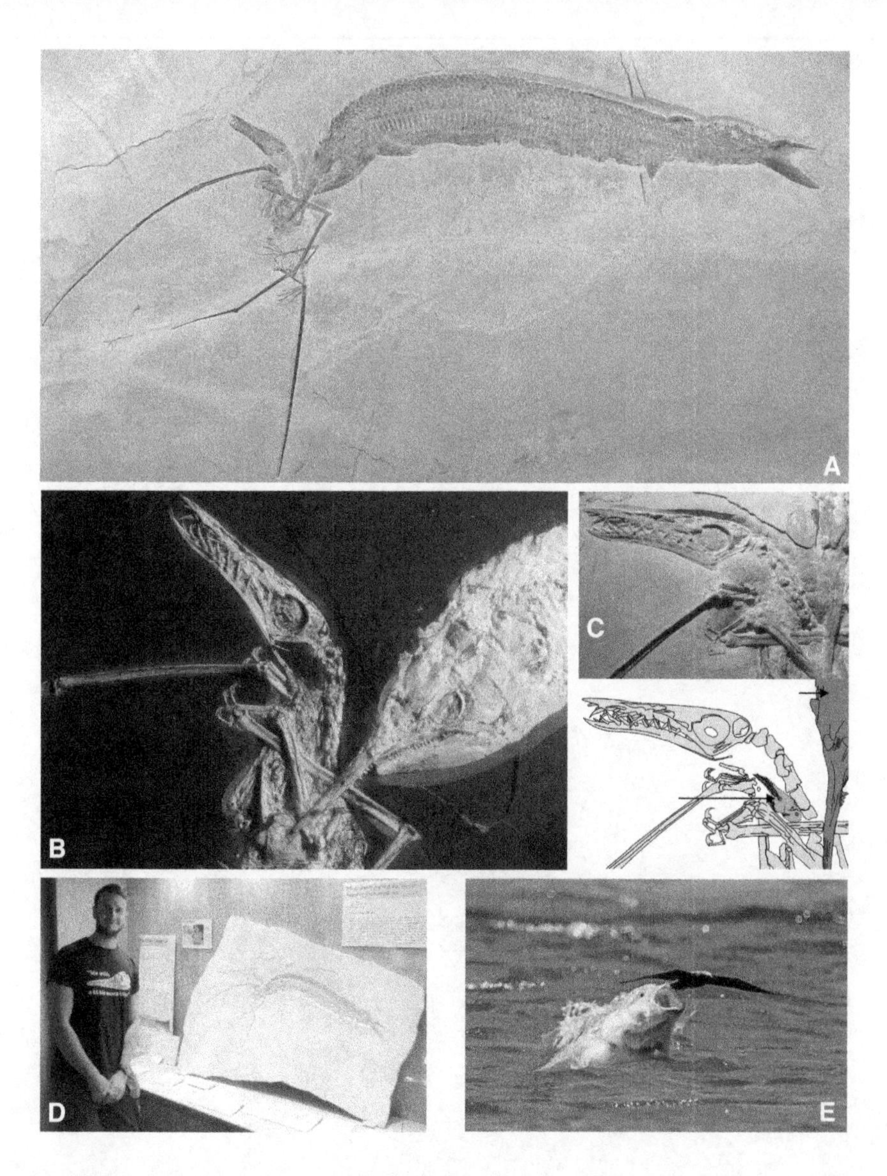

FIGURE 4.5. (*A*) The pterosaur *Rhamphorhynchus muensteri* snagged by the predatory fish, *Aspidorhynchus acutirostris*. (*B*) Detail under UV light showing position of the upper jaw of *Aspidorhynchus* caught in what would have been the pterosaur's wing. (*C*) Close-up of *Rhamphorhynchus* skull with sketch showing where the tiny, fragmentary bones of the fish are preserved in the gullet. Upper arrow points to the head of the large fish, and lower arrow points to the fragmentary fish inside the pterosaur's gullet. (*D*) The author with the "Jurassic Drama." (*E*) Modern-day giant trevally ambushing a bird as part of David Attenborough's *Blue Planet II*.

([*A*, *C*] Courtesy of Helmut Tischlinger, illustration by Dino Frey; [*B*] courtesy of Mike Eklund and the Wyoming Dinosaur Center; [*D*] courtesy of Levi Shinkle; [*E*] courtesy of BBC Studios)

FIGURE 4.6. (Overleaf). *Their final ambush.*

Diving into the water, a *Rhamphorhynchus* is quick to snatch a tiny fish at the exact moment that a hungry *Aspidorhynchus* goes in for the kill.

BOB NICHOLLS 2020

the pterosaur, or if it was actually hunting the flying reptile. If the latter, did it catch it under the water or jump out as the pterosaur swooped down for the fish? The presence of the small fish inside the pterosaur's gullet suggests that the coming together happened during or immediately after the pterosaur completed a successful hunt. In the wild, when watching living animals, lucky photographers and filmmakers occasionally capture multiple feeding interactions like this all at once, usually after following an intense predator–prey chase that ends with the prey being caught, only for both to be eaten by a larger predator. Every so often, bizarre circumstances are snapped on camera when animals accidently become tangled together, occasionally during a hunt, such as a billfish found with its pointed bill embedded in another animal, or a bird with its beak pierced straight through the body of a fish.

So far, no *Rhamphorhynchus* remains have been found inside *Aspidorhynchus*, but that does not mean this was a single freak incident. Instead, it seems that the fish misjudged the size of its pterosaur prey, because its mouth gape was too small to swallow this individual whole. Or it simply might have been an accident. Intriguingly, a fossilized regurgitate was found to include the partly digested vertebrae and finger bones of a *Rhamphorhynchus*, and although it is impossible to say who ate this pterosaur, a predatory fish is the prime suspect. Perhaps it was from an *Aspidorhynchus*?

Emphasizing that this association is not just a single oddity, four other nearly identical fossils are known. In each case, the *Rhamphorhynchus* is entangled with the jaws of *Aspidorhynchus*, and so this relationship is not coincidental. Rather, this predatory fish seems to have hunted the pterosaur, perhaps exploiting the situation when the pterosaur was immersed under the water, itself on the hunt. These failed attacks indicate lethal errors in judgment by the *Aspidorhynchus*. In comparison, some modern-day fish feed on birds, and some even jump out of the water to catch seabirds, like the giant trevally that was filmed vividly doing so in David Attenborough's *Blue Planet II* documentary.

The fact that five exquisite examples of *Rhamphorhynchus* and *Aspidorhynchus* have been found together strongly suggests that this was a common, albeit unusual and perhaps unexpected, predator–prey relationship that existed between two very different species during the Jurassic. The one-of-a-kind specimen found in 2009 captures a rare fossilized glimpse of an extraordinary food chain in action.

TERROR WORM OF THE PRIMEVAL SEA

Dinosaurs, saber-toothed cats, and Megalodon are among the animals that come to mind when we think of prehistoric predators. With such a vast array of fearsome fossil brutes, admired for their often giant size, deadly toothy weapons, or top predator status, smaller predators are easily forgotten and their stories lost in time.

Even though such iconic predators take center stage, species that lived long before any of these animals had even evolved were top predators in their own world and had already become fossils prior to the appearance of dinosaurs and their contemporaries. The earliest examples are recorded from the Cambrian, when life exploded onto the scene and animals began to eat each other, forming the first complex predator–prey relationships more than half a billion years ago.

One of the most extraordinary insights we have for such a relationship is captured in a predatory priapulid-like marine worm called *Ottoia*, whose fabulous and abundant fossils come from the famous Burgess Shale in British Columbia (see chapter 2). Priapulids are a group of seabed-burrowing carnivorous worms, commonly known as "penis worms" in reference to their supposedly phallic body shapes, and are represented today by around twenty species. Despite the substantial separation in time, the exceptionally preserved soft-bodied fossils show that *Ottoia* roughly resembles its modern-day cousins, both in body shape and in having a retractable spiny feeding proboscis.

A 505-million-year-old spiny penis worm might not sound terrifying to you (or perhaps the image it conjures does), but this species was larger than many of the tiny animals in its ecosystem. *Ottoia* reached a maximum length of around 15 centimeters and played an important role in the food web.

Though it was not the deadliest predator of its time, actively cruising around and hunting prey, it was among the largest in the muddy seabed, where it lived in underground burrows. Sitting in wait, *Ottoia* was probably an ambush predator that relied on chemical cues to trigger an attack. Any animal that wandered too close to its burrow was in for a shock. *Ottoia* had a menacing-looking proboscis surrounded by multiple rows of hooks and spines and ending in a mouth lined with sharp pharyngeal (throat) teeth, which it used to grasp prey.

Ottoia was the first Cambrian critter to show direct evidence of a predator–prey relationship in a Cambrian ecosystem, judging from specimens with their preserved gut contents. The pioneering work on *Ottoia* was published in 1977, and substantial evidence of its diet has since been found, epitomized in an extensive study in 2012. During this research, a whopping 2,632 *Ottoia* specimens were analyzed for the presence of gut contents. The Burgess Shale material is so exceptional that soft parts of an animal's body that otherwise would not preserve can be seen, such as internal organs. As a result, the gut of *Ottoia* is often traceable from the pharynx to the anus.

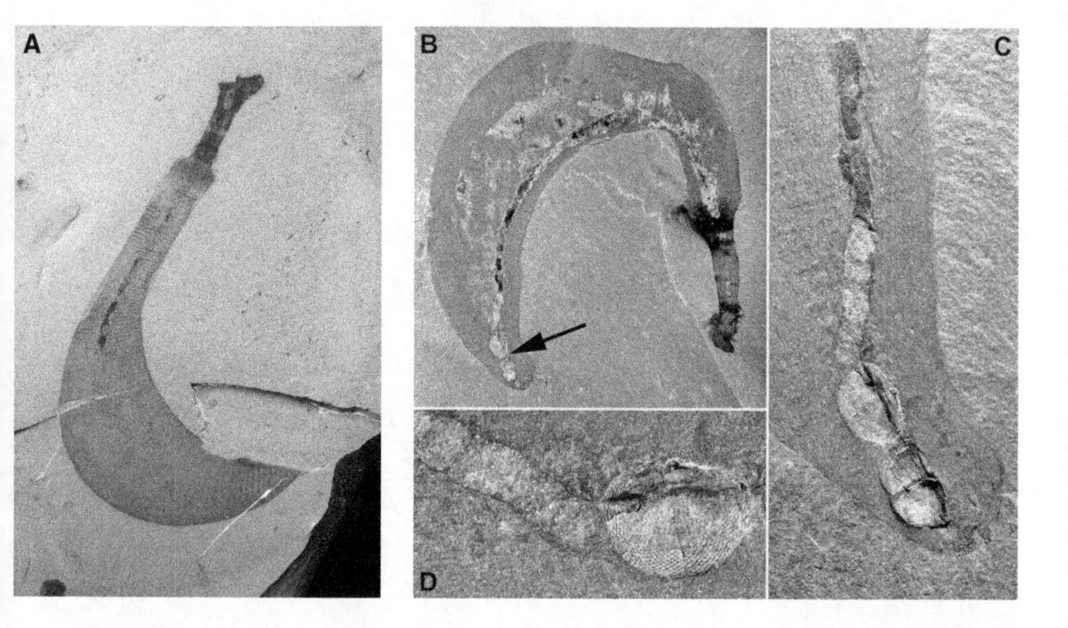

FIGURE 4.7. (*A*) Complete specimen of *Ottoia* with extended proboscis. (*B*) An example of *Ottoia* perfectly showing the outline of the gut, with arrow pointing to the last meal. (*C–D*) Two brachiopods contained in the gut contents of the same specimen.

([*A*] Courtesy of Wikimedia Commons, Martin R. Smith; [*B–D*] courtesy of Jean Vannier)

BOB NICHOLLS 2020

The gut was identified in 2,000 specimens, with the last meal preserved in a staggering 561 individuals, 21 percent of the total sample size. The remains show that *Ottoia* dined on a wide range of contemporary species, some of which were often swallowed whole. This includes shelled invertebrates (hyolithids and brachiopods), different types of tiny arthropods (trilobites, agnostids, and bradoriids), spiny animals called wiwaxiids, and a polychaete worm. In some cases, multiple examples of each species or a combination of different species were also found inside the gut. Some evidence suggests that *Ottoia* even might have been cannibalistic, as one specimen was found associated with the gut contents of another. However, it is not possible to say whether it was preserved inside the gut or was lying on top of it. In three exceptional specimens, *Ottoia* is directly associated with a carcass of the similarly sized arthropod called *Sidneyia*; one of these contains at least five individuals representing both juveniles and adults. Apparently, they were in the process of feeding, indicating that *Ottoia* was also an active scavenger.

These findings suggest that *Ottoia* was a prolific predator. Although hyolithids were its preferred snack, being most common inside the gut contents, the variety of species (as many as nine) shows that it was a dietary generalist and would ingest whatever food it could find, by either hunting or scavenging. Support for this idea comes from some recent priapulid species (such as *Priapulus caudatus*) that feed on a variety of animals, implying a comparable feeding type.

Just as ecosystems today are greatly influenced by the crucial role that predators play, it is mind-boggling to think that the same behavioral interaction of one species killing and consuming another has its origins buried so deep in time that it can be reliably reconstructed from fossils as ancient as *Ottoia*.

FIGURE 4.8. (Opposite). *The early worm catches . . .*

Various tiny animals are ambushed and attempt to flee from the "giant," menacing *Ottoia*, which has caught a shelled hyolithid.

GREEDY FISH

In spring 1952, famed fossil hunter George F. Sternberg was hosting a fossil fish hunt in Gove County, Kansas, for Bob Schaeffer and Walter Sorenson, two paleontologists from the American Museum of Natural History (AMNH) in New York City. While searching for fossils, Sorenson found a large flattened bone that appeared to be part of a tail from what looked like a large fish. Sternberg was quick to confirm and further identified it as being from the large predatory fish *Xiphactinus audax*. This gnarly looking fish somewhat resembles a modern tarpon with its upturned, bulldog-like mouth, but the jaws were packed full of protrusive, fang-like teeth. It was a fast, powerful swimmer and among the top predators of its day.

Before the AMNH team had to leave and continue their travels, the trio uncovered a bit more of the tail, chipping it out from the overlying chalk rock that entombed it. As there were no guarantees of its completeness, a lack of necessary funds to ship it back to New York, and the fact that the AMNH had already procured one of these Kansas fossil fish, they kindly let Sternberg collect and keep it (he lived locally).

A couple of weeks later, on June 1, Sternberg returned to the site and began excavating the fossil. Fearing that the specimen would be damaged or lost, he camped next to the fish under the hot Kansas sun until it was fully uncovered later that month. His hard work and patience paid off. The fish turned out to be the finest known example of *Xiphactinus*. Measuring roughly 14 feet long, this was a big fish. By comparison, the largest specimen found to date is 18.5 feet long and was collected in 2008. Sternberg contacted Schaeffer at the AMNH to explain how excellent the fish had turned out and even offered the specimen to him, but Schaeffer declined his kind gesture because it was Sternberg who had done all of the rigorous work.

At the time of its discovery, Sternberg's fish was the most complete and best *Xiphactinus* in the world. Yet, there was something else that sent paleontologists and the public into a bit of a frenzy. Having exposed the fish completely in the ground, while still out in the field, Sternberg began removing bits of chalk that overlay some of the bones. In doing so, he made a shocking discovery.

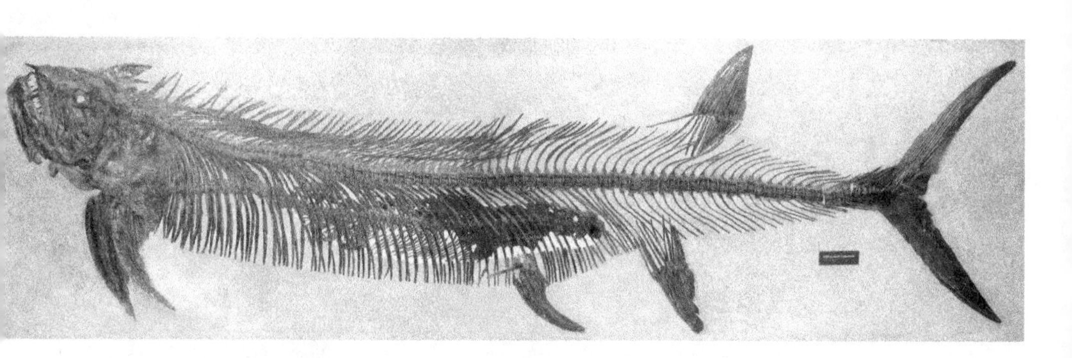

FIGURE 4.9. Sternberg's original "impossible fossil" of a *Xiphactinus audax* with a *Gillicus arcuatus* inside—the famous "fish-within-a-fish."

(Courtesy of Mike Everhart)

FIGURE 4.10. Historical photograph of George F. Sternberg (*left*) carefully cleaning his perfect fossil find in 1952.

(Courtesy of the George Sternberg Photographs Collection, University Archives, Fort Hays State University)

Preserved between the ribcage was another complete fish with its head facing in the opposite direction to that of *Xiphactinus*. Was it an unborn fetus? Definitely not. This second fish is pretty large, just over 6 feet, and belongs to a totally different species, called *Gillicus arcuatus*. Sternberg had landed the ultimate catch, a *Xiphactinus* and its final meal, direct evidence of predator and prey. This fish-within-a-fish was referred to as Sternberg's "impossible fossil" and became one of the most recognizable and photographed fossils in the world. Today the specimen is displayed in pride of place at the Sternberg Museum of Natural History.

Xiphactinus clearly swallowed its *Gillicus* prey whole head first, but it was just too large for the hunter to swallow comfortably. The tail appears to be lodged in (or near) the throat region, and there is nothing to suggest that the digestion process had begun (no acid etching on the bones). As a result, not only is it probable that the *Xiphactinus* choked on its prey, but *Gillicus* quite possibly injured it. Struggling and thrashing around in an attempt to escape, its sharp fin spines may have punctured the esophagus or stomach (or maybe even ruptured vital organs) and doomed them both to a watery grave at the bottom of the Western Interior Sea. This all might have happened in a matter of minutes.

Many animals swallow other animals whole as part of their hunting behavior. Snakes are probably the best known for perfecting this technique, because they do not chew, though even snakes can have eyes bigger than their bellies. When a snake swallows a meal that is too large, the animal will sometimes get stuck or cause internal damage, which can lead to the snake's death. Just like *Xiphactinus*, predatory fish often swallow their prey whole, and larger prey can become stuck more easily. For instance, a dead great white shark was found in 2019 with an oversized and very dead turtle trapped in its mouth. In some cases, even if swallowed whole, the prey can fight back and escape, irritating the predator enough that it is regurgitated, or the eaten becomes the eater (or the excavator) and will dig its way out. In one odd example, a blind

FIGURE 4.11. (Opposite). *The ill-fated optimist.*

A *Xiphactinus* biting off more than it could chew: a whole *Gillicus*, which is about to get lodged in its throat.

snake was eaten by a toad but lived long enough that it found its way through the toad's gastrointestinal system and out the cloaca!

This fish-within-a-fish is not a one-off discovery. Several *Xiphactinus* fossils have been found with *Gillicus* as their last meal, including yet another with an entire *Gillicus* inside. A third complete *Xiphactinus* was identified with the skeleton of a different type of fish between its ribs, called *Thryptodus*. Numerous other *Xiphactinus* have been found with the partially digested and fragmentary bones (usually vertebrae) of *Gillicus* in their stomach cavity; it obviously had a particular taste for *Gillicus*.

Such exceptional predator–prey fossils capture evidence of typical hunting behavior in action. Unfortunately for these toothy individuals, some 85 million years ago they made a bad dinner choice, biting off much more than they could chew.

CRACKING THE CASE OF THE BONE-CRUSHING DOGS

Figuring out how and what a prehistoric predator ate when its last meal is not preserved is hard. It can be frustrating when paleontologists can never truly verify whether any of their hypotheses are correct. In some situations, by drawing from several lines of evidence, it is possible to deduce with a heightened level of certainty that a particular prehistoric species ate a certain type of food in a specific way.

An excellent example is found in the extinct subfamily of dogs (canids) called Borophaginae, better known by their awesome nickname, "bone-crushing dogs." Borophagines were common in North America and flourished for more than 30 million years, before the last of them died off just 2 million years ago, right before the beginning of the Ice Age.

Despite their name, not all borophagines were capable of crushing bones. Many of the earlier forms were small, fox-sized omnivores, whereas the larger bone-crushers appeared later and were apex predators. The massive *Epicyon haydeni* was as big as a brown bear, weighed as much as a lion, and is the largest-known canid of all time. These big dogs had robust skulls and jaws, domed foreheads, very strong teeth, and a powerful bite, as inferred by large muscle attachment areas. Detailed computational models have shown that their enlarged cheek teeth were able to deal with high levels of strain, which is consistent with fossils that display excessive tooth wear from the cracking of bones. Similar features and findings are shared today only by hyenas in Africa and Asia, which are famous for cracking open and eating the bones of their prey. Does this suggest that bone-cracking borophagines did the same thing?

These ancient dogs seemingly have features that allow us to infer their feeding and dietary habits, backed by comparisons with their modern-day hyena analogues, and computer simulations showing that they had tremendously strong teeth capable of withstanding intense pressure. Hence, for a very long time, borophagines have been viewed as the North American ecological equivalents of hyenas, and it has been assumed that they, too, cracked open and swallowed bones, although direct evidence has been absent.

The last surviving member of the Borophaginae was *Borophagus*, which the subfamily is named after. Compared with all the dogs, its premolar teeth were among the most massive, best developed, and specialized for cracking bones. Weighing as much as 40 kilograms or more, it can be imagined as a powerful, wolf-sized dog with hyena-like bone-cracking dentition. Its fossils are particularly common and have been found at several sites across North America, but their bones do not provide the answers. Rather, the proof was in their poop.

To know exactly what an animal has been eating, you need to look at what comes out the other end. It seems obvious, but this is far more difficult when working with extinct species. Although countless coprolites (fossil feces) have been found, it is extremely rare to find a fossilized animal associated with fecal remains and thus work out who dung it. It is, however, possible to confidently link a coprolite with its producer when they are found in the same rocks. This is what paleontologist and fossil dog expert Xiaoming Wang of the Los Angeles County Museum of Natural History and colleagues did while studying a new collection of rare coprolites from Stanislaus County, California. The fossil feces are from a 5.3- to 6.4 million-year-old Miocene rock formation known as the Mehrten, where *Borophagus* remains are found in abundance.

Lo and behold, the coprolites are packed full of bone. This strongly suggests that they were produced by *Borophagus*, the only large-sized carnivore from that formation that could possibly have made the prehistoric scat, which is comparable in size to living wolf scat. A micro-CT scan revealed that the undissolved bone fragments were preserved both on the surface and within the coprolites and are often rounded, polished, or acid-etched. Some pieces could be identified, including a bird limb bone, a fragment of jaw from a beaver, a piece of skull from a medium-sized mammal, and part of a rib from a large, deer-sized mammal. The scat confirms that *Borophagus* fed on a variety of animals and routinely consumed their bones, perhaps even the entire carcass, as spotted hyenas are known to do. However, these bony fragments in the ancient poop suggest that although *Borophagus* consumed bones as spotted hyenas do, it digested them more in the way of striped hyenas and brown hyenas.

The droppings were found in clusters reminiscent of scent-marking behavior, similar to latrine areas used by many living social carnivores. *Borophagus* probably lived in social groups that, as evidenced from the

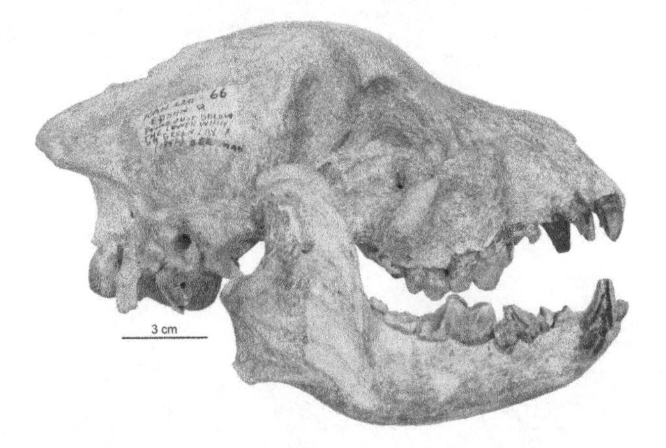

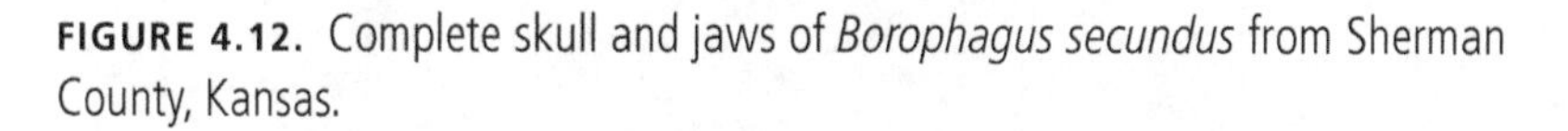

FIGURE 4.12. Complete skull and jaws of *Borophagus secundus* from Sherman County, Kansas.

(Image from Wang, X., et al. 2018. "First Bone-Cracking Dog Coprolites Provide New Insight Into Bone Consumption in *Borophagus* and Their Unique Ecological Niche." *eLife* 7, e34773)

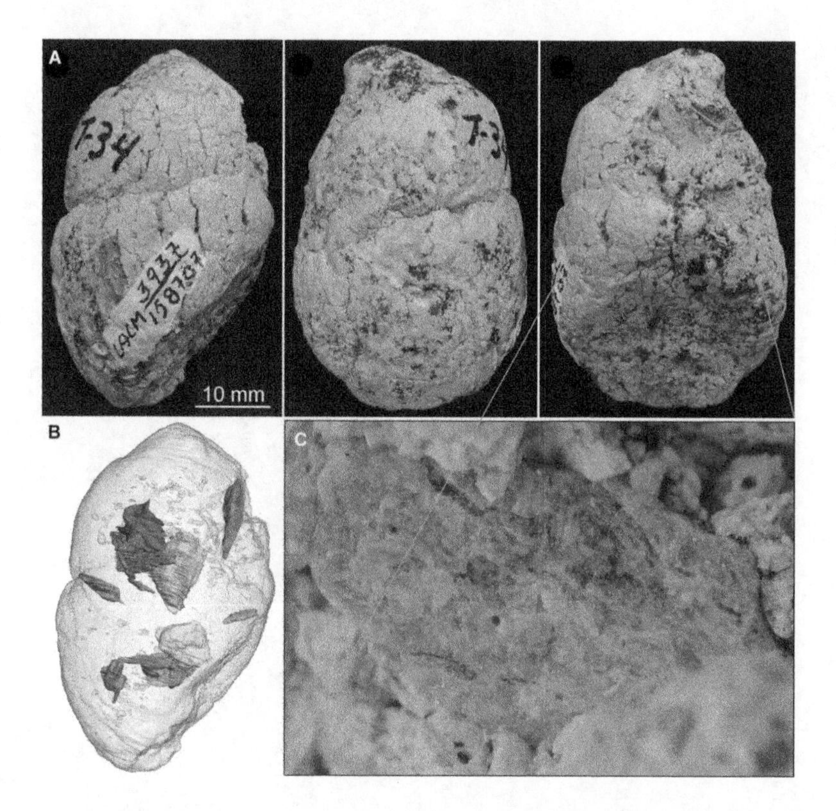

FIGURE 4.13. (*A*) Multiple views of a coprolite assigned to *Borophagus*. (*B*) Internal contents of same coprolite showing multiple bone fragments. (*C*) Close-up of a bone fragment embedded in the coprolite.

(Images from Wang, X., et al. 2018. "First Bone-Cracking Dog Coprolites Provide New Insight Into Bone Consumption in *Borophagus* and Their Unique Ecological Niche." *eLife* 7, e34773)

BOB NICHOLLS 2020

coprolites, were capable of taking down prey much larger than themselves, as do spotted hyenas and wolves. Besides hunting in packs, *Borophagus* probably out-muscled other predators, stealing their kills, and occasionally scavenging free meals by cracking open (and consuming) leftover bones to obtain the nutritious, calorie-dense marrow inside.

Borophagus and gang independently evolved a highly specialized bone-cracking behavior akin to extinct and modern-day hyenas. Although bones have yet to be found inside the gut of one of these dogs, their fossilized scat has unambiguously confirmed the long-held suspicion that these top predators consumed the bones of their prey and filled an ecological niche that no longer exists in present-day North America.

FIGURE 4.14. (Opposite). *The bone processor.*

A *Borophagus secundus* gracefully taking a poop next to several dried scats, which are white due to the presence of calcium from bones. Other members of the *Borophagus* pack can be seen resting in the background; one is crushing a bone.

TO CATCH A KILLER: THE SNAKE THAT DINED ON BABY DINOSAURS

Fossils are usually found encased in rock with only a small portion exposed, perhaps a couple of bones or a piece of shell. As is the nature of fossils, not all specimens are complete; in fact, most tend to be fragmentary, so often it is a bit of a gamble as to whether the animal might be completely preserved, concealed inside the rock. Paleontologists tend to have the ability, almost like a superpower, to identify how complete or significant a specimen might be simply by eyeballing the partly exposed fossil. This means that if the specimen displays nothing unusual or particularly rare, it can be put to the back of the queue and reexamined or cleaned in the laboratory at a later date. Sometimes, however, things can be overlooked, and it can take years before their significance is realized. Often, these (re)discoveries are made by paleontologists when routinely browsing through fossil collections.

As part of a field expedition in 1984, near the village of Dholi Dungri, western India, paleontologist Dhananjay Mohabey collected a series of large blocks of rock that contained three 67-million-year-old dinosaur eggs. Dinosaur eggs are common in the area and are found at several locations around the world. Based on their spherical shape and large size—16 centimeters in diameter—it was clear that these eggs were laid by a sauropod dinosaur, the same group of long-necked, long-tailed giants that include *Diplodocus* and *Brontosaurus*. Sauropod eggs are typically spherical and large, sort of watermelon-shaped, so it was pretty clear that Mohabey's eggs belonged to a sauropod.

One of these eggs was crushed, but the other two were intact and unhatched. Judging from their close association, compared with other clutches of dinosaur eggs found in the same area, all three belonged to part of a clutch that typically would have contained six to twelve eggs. Preserved adjacent to the crushed egg were the delicate bones of a hatchling sauropod that had just emerged from its egg. This tiny sauropod was an important find, given that fossilized sauropod hatchlings are rare. Yet, the specimen's most unusual story was still to be realized.

A couple of years after the discovery, it turned out there had been a case of mistaken identity. Vertebrae thought to belong to the sauropod hatchling

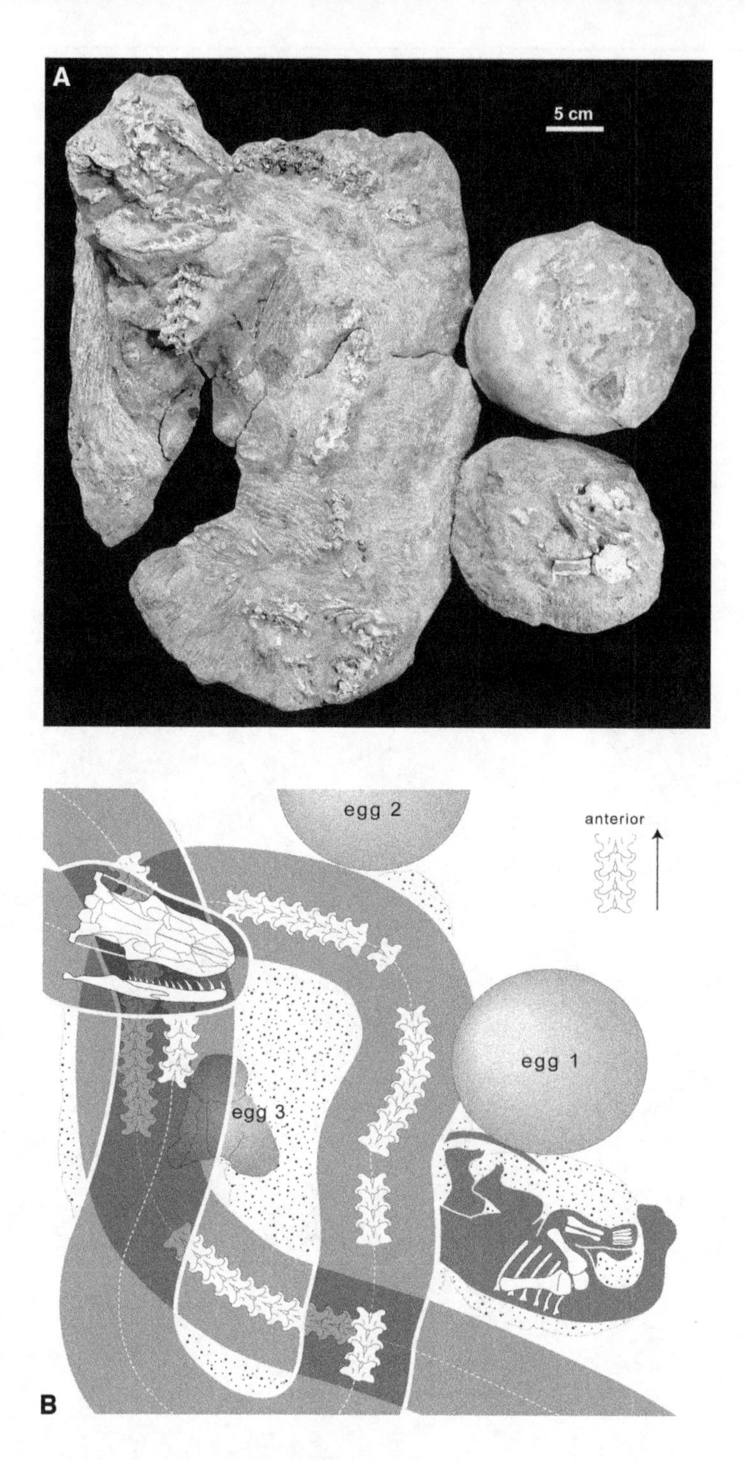

FIGURE 4.15. (*A*) The snake, *Sanajeh indicus*, coiled around the eggs of a sauropod with the fragmentary remains of a tiny newborn adjacent to one of the eggs. (*B*) Interpretive illustration showing position and identification of the skeletal and egg remains.

(Images from Wilson, J. A., et al. 2010. "Predation Upon Hatchling Dinosaurs by a New Snake from the Late Cretaceous of India." *PLOS One* 8, e1000322)

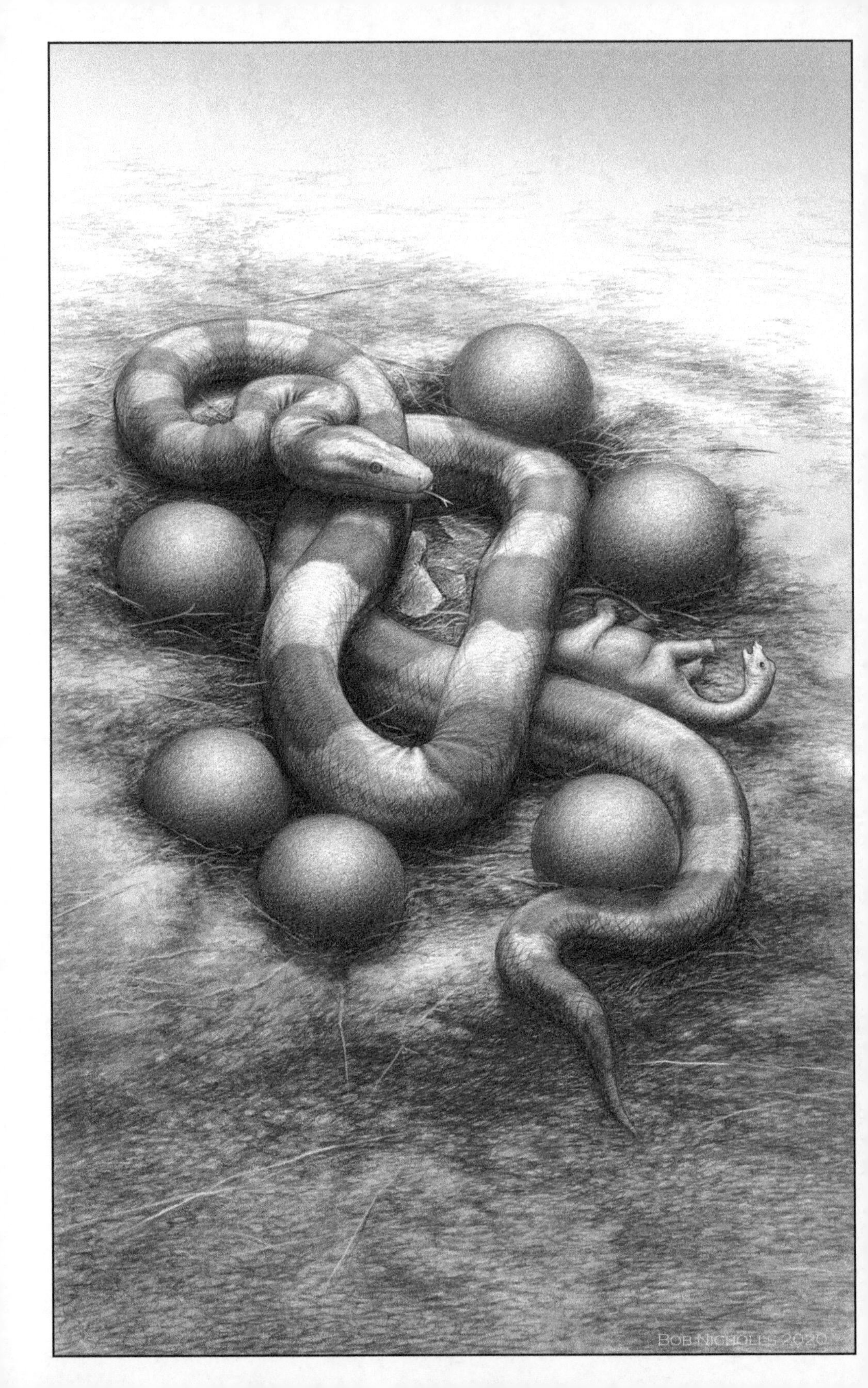
BOB NICHOLLS 2020

were actually from a snake. This was confirmed in 2001 by paleontologist Jeffrey Wilson, who had visited India and studied this specimen with Mohabey. Thinking that more of the snake might be buried inside the entire block, it was agreed that the specimen would be transported to the University of Michigan temporarily, where the rock surrounding the vertebrae and eggs could be carefully removed.

To the team's amazement, the removal of the rock revealed an almost complete snake sitting in the nest, surrounding the three dinosaur eggs and hatchling. Snakes from the Cretaceous are rare, and the specimen was identified as a new species, called *Sanajeh indicus*. The occurrence of a snake in a sauropod dinosaur nest raised several questions regarding the feeding ecology of this prehistoric serpent. Namely, was it feeding on the eggs? If so, how?

Many living species of snake eat eggs, often whole—an impressive yet bizarre feat. Given the association between the snake and eggs, it could be inferred that *Sanajeh* swallowed these particular eggs whole. That seems to make sense, except that features of the skull and skeleton show that *Sanajeh* lacked the specialized wide gape of modern egg-eating snakes. Thus, it was thought to be incapable of swallowing an entire large sauropod egg, the size of which greatly exceeds the snake's gape.

Instead, it is plausible that *Sanajeh* lived a lifestyle similar to that of the living Mexican python snake, which breaks the eggs of the Olive Ridley sea turtle by constriction and then consumes the contents. This is the most likely feeding scenario, although it is also possible that the snake simply waited for the baby sauropod to do the hard work—crack out of the egg—so it could then make its move. In all probability, the snake employed a combination of both strategies. Providing additional evidence for this predator–prey interaction is the fact that other *Sanajeh* bones have been found alongside sauropod eggs in the same area, suggesting that these snakes dined on sauropod hatchlings in a typical feeding behavior.

FIGURE 4.16. (Opposite). *Sanajeh in the nest.*

Slithering into the nest of a giant titanosaur sauropod, a *Sanajeh* snake is coiled around eggs and is readying to strike a newborn hatchling.

At 3.5 meters long, *Sanajeh* maintained a huge body size advantage over the small, 50-centimeter-long hatchlings, leaving them highly vulnerable to attack. Their only form of defense was to grow quickly. Adult sauropods were longer than two or three buses and were clearly off the snake's menu.

During the late Cretaceous period, 67 million years ago, a baby dinosaur hatched from its egg and took its first breath in a new and alien world. Finding its feet, the hatchling stumbled from the egg for a couple of steps. Attracted by the scent and movement of the newborn, *Sanajeh* entered the nest, coiled its body around one of the eggs in a clockwise fashion, with its head resting on the topmost loop of its body, and readied to strike. At that critical moment, the predator and would-be prey were buried alive, smothered and entombed for all eternity by a storm-induced sandy mudslide. We know, based on the rock the specimen is contained in, along with the local geology of the site, that the prehistoric environment would have been subtropical, with periods of wet and dry, resulting in heavy storms that sometimes would have led to mudslides.

Instead of feasting on its next meal, this snake was caught dead in the act, covered during the flow of a mudslide that led to a quick and instantaneously deep burial. You can only hope that the hatchling was unaware of what was about to happen. This association provides evidence that snakes once dined on dinosaurs, feasting on babies of the largest animals that ever walked on Earth, something we only know because of this most exceptional fossil.

A DINOSAUR-EATING MAMMAL

Dinosaurs usually steal the headlines when it comes to prehistoric predators. Although these reptiles reigned supreme for millions of years, it can be easy to forget that our earliest mammal ancestors were there, too, scurrying in the shadows. These typically mouse-sized mammals had to avoid becoming prey for those quick enough to catch them. Undoubtedly, in a world dominated by dinosaurs, the furry critters would have been on the menu for many theropods, big and small, but what if the tables were turned?

Contrary to popular belief, not all dinosaur-era mammals were tiny, as was revealed by the major finding of two new species of fossil mammals collected from the rich Cretaceous rocks in Liaoning Province, China. The first was *Repenomamus robustus*, described in 2000, and then the larger, aptly named *R. giganticus* was brought to the attention of scientists in 2005.

When we see words like *robust* or *giant* used with *fossil*, we immediately think of something enormous, perhaps in this case picturing an ancient mammal the size of a *Diplodocus* or a *Stegosaurus*. Not even close. Adult *R. robustus* were roughly the size of a cat, whereas *R. giganticus* was about as big as a large badger, maxing out at just over a meter long and weighing around 12–14 kilograms. That might appear relatively small, but putting this into perspective, in the grand scheme of what we know about early mammal evolution, *R. giganticus* really is a giant. By comparison, many small Mesozoic mammals have tiny skulls, measuring just 1–5 centimeters long, and are thought to have been nocturnal insectivores, whereas the skull of *R. giganticus* is 16 centimeters—obviously a major difference. As such, *Repenomamus* is the largest known Mesozoic mammal identified from substantially complete skeletons.

Its large body size, coupled with a robust skull sporting strong, pointed teeth, would suggest that *Repenomamus* was a predator. It probably successfully competed with the dinosaurs that lived alongside it, especially considering that many of the smaller-bodied feathered theropods (including dromaeosaurs) weighed only a few kilograms. So, what did this brute eat? Insects, fish, or smaller mammals, maybe? Although that seems likely, one thing we know for certain is that it ate dinosaurs—specifically, baby dinosaurs.

When one of the *R. robustus* fossils was collected, only a couple of bones were exposed, so it was taken to the laboratory for further assessment. As the rock was removed during preparation, it turned out to be an almost complete specimen, but that was not all. The careful work had uncovered an association of jumbled bones and teeth lodged between the ribcage, in the same spot where the stomach is positioned in living mammals. This was the remains of its last meal. On close inspection, the tiny serrated teeth, limbs, and fingers were found to match those of juvenile *Psittacosaurus*, a ceratopsian dinosaur (see chapter 2). Coincidentally, adults and juveniles of this herbivore are common in the same rocks in which *Repenomamus* was found.

The partly digested juvenile has an estimated length of just 14 centimeters, and its teeth show wear from eating, indicating that it was certainly not an embryo that had been taken from an egg. Some of the long limb bones remain in articulation, but the skull is broken into pieces, as are other bones, suggesting that *Repenomamus* dismembered this juvenile before swallowing it in bite-sized chunks, similar to how crocodiles feed. We know from several *Psittacosaurus* accumulations that juveniles of this age tended to stay together in groups, usually near their parents, so it is possible that *Repenomamus* snatched hatchlings while they wandered away, or perhaps it actively raided their nests. It seems unlikely, however, that *Repenomamus* would have challenged adult *Psittacosaurus* because they were twice their size.

Instead of being presented with direct evidence of dinosaurs dining on early mammals, which fits our preconceived ideas of what we *should* expect—and what certainly must have happened—we are instead faced with something extraordinary. The presence of baby dinosaur bones inside the gut of this 125-million-year-old mammal opens up an entirely different world, one that we would never have envisaged. *Repenomamus* might have been imagined as nothing more than furry theropod food if it were not for this amazing fossil.

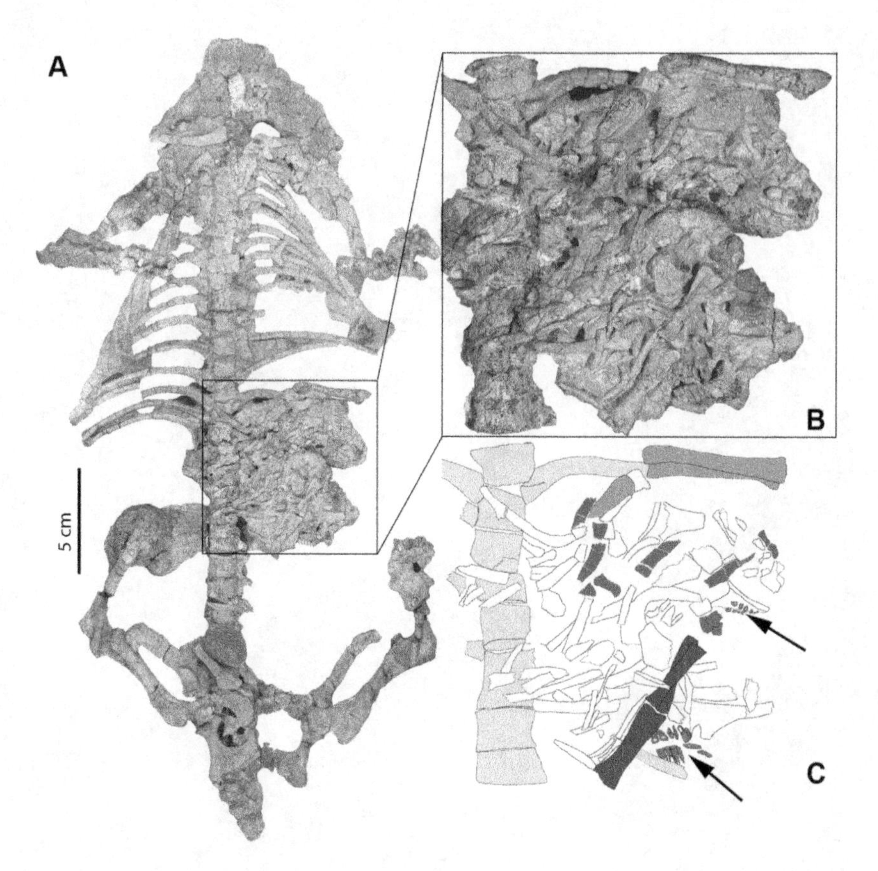

FIGURE 4.17. (*A*) The skeleton of *Repenomamus robustus* containing the incomplete and partly digested skeleton of a baby *Psittacosaurus* inside its stomach. (*B*) Close-up of the stomach contents. (*C*) Illustration of the stomach contents containing the fragmentary, ripped-up remains of the baby. Arrows point to the teeth of the dinosaur.

(Courtesy of Jin Meng)

FIGURE 4.18. (Overleaf). *Happy Mr. Whitesocks.*

A lone *Repenomamus* proudly shows off its kill, a stray baby *Psittacosaurus*, while in the background, a subadult *Psittacosaurus* and multiple babies walk away through the dense undergrowth.

BOB NICHOLLS 2020

THE FEEDING GROUND: "SOMETHING INTERESTING?"

Wyoming is home to some of the world's most famous dinosaurs. To name but a few, *Diplodocus*, *Stegosaurus*, and *Allosaurus* have been found here, collected from the renowned and dinosaur-rich Late Jurassic Morrison Formation. An abundance of dinosaur bones from this formation was discovered in 1993 on the Warm Springs Ranch in the small town of Thermopolis, in the Bighorn Basin. This find was enough to warrant the creation of a new museum—the Wyoming Dinosaur Center (WDC).

The museum opened in 1995, and rather serendipitously, one of the most significant discoveries was made in the same year. Excavation of a newly discovered quarry in the 150-million-year-old rocks, revealed various dinosaur tracks, along with dozens of theropod (*Allosaurus*) teeth alongside scattered sauropod (*Camarasaurus*) bones. Finding dinosaur body and trace fossils like this in the same rocks is incredibly rare. The geology of the quarry shows that the *Camarasaurus* lies at the edge of an ancient lake, and to preserve all of these associations in context, a shelter was built to protect it. In light of the intriguing nature of the site, it was called "SI," short for "Something Interesting." But what exactly was so interesting?

Camarasaurus grew to around 18 meters long and weighed about 15 tons and is the most common dinosaur found at the WDC quarries. About 40–50 percent of the skeleton at SI has been uncovered and represents a juvenile, roughly half the size of the largest known individuals. One of the most fascinating and rare findings at the site was a large depression surrounding parts of the juvenile's legs, hips, and tail bones. This represents the outline of where the body was originally laid in the mud at the edge of the shallow-water lake. The ancient mud shows heavy trampling from the passage of many sauropods (which paleontologists call *dinoturbation*), and some of the *Camarasaurus* bones were stepped on and crushed.

Since the initial discovery, more than 150 *Allosaurus* teeth and many of their three-toed tracks, including distinct scratch marks, have been found dotted around the outline and bones of the *Camarasaurus*. Some of the bones bear claw marks, and others display tooth damage matching the *Allosaurus* teeth. The presence of many shed teeth from young juveniles and adults shows that multiple allosaurs, perhaps a group or family, were feeding on the sauropod and lost their teeth in the process. The scattered nature of the sauropod's skeleton shows that the carcass was literally torn apart; even stomach stones were exposed from the sauropod's insides. This was an *Allosaurus* feeding frenzy.

At a maximum of 10 meters or so, *Allosaurus* was the apex predator of its day. A full-grown adult could possibly have taken down a juvenile *Camarasaurus* with ease, especially if it had help. However, there is nothing to suggest that one or more *Allosaurus* actually made the kill; it seems more likely that the allosaurs were attracted by a decaying carcass, which they scavenged. No allosaur bones have been found at SI, suggesting that either the allosaurs arrived at the feeding ground one after the other or that they tolerated the presence of one another and thus might have been a social group.

On a personal level, the WDC and the SI quarry hold a special place in my heart, because I spent many months there digging up dinosaurs and studying the collections. My first day in 2008 was like something out of an adventure movie. Although this seems rather like a cliché, while listening to the *Jurassic Park* soundtrack, I traveled by jeep across rugged terrain and up vast hills to the SI quarry. Over the years, I have taken part in several digs there (and at many of their other quarries), unearthing *Camarasaurus* bones and *Allosaurus* teeth, including co-finding (with Levi Shinkle) one of the largest *Allosaurus* teeth ever discovered at SI. To this day, SI continues to be excavated and new finds are made, helping to reveal more about the extraordinary story of this Jurassic feast, which was most *definitely* interesting.

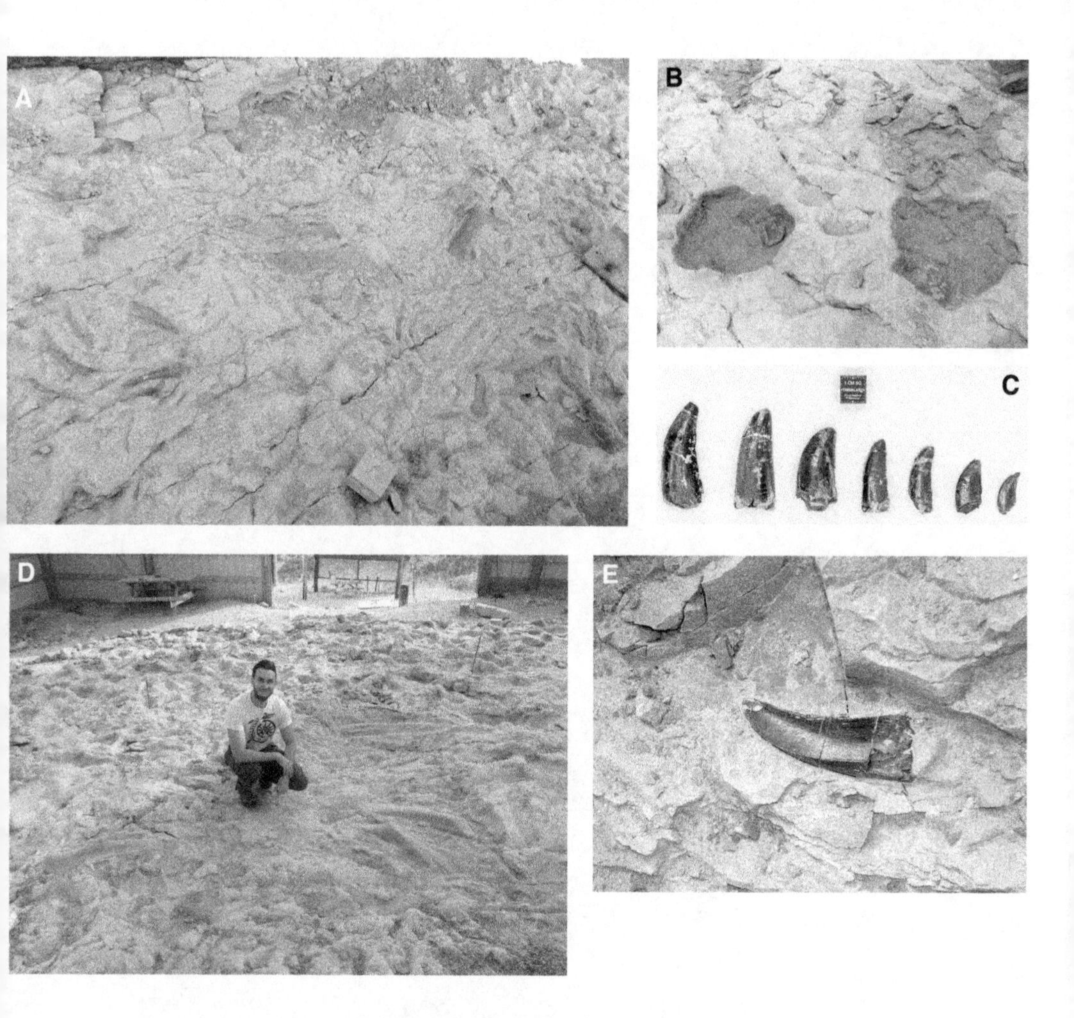

FIGURE 4.19. The SI ("Something Interesting") dig site. (*A*) Various bones of the juvenile *Camarasaurus* surround a large depression where the body was laid; note the three distinct claw marks to the left. (*B*) Large sauropod footprints. The left contains a squished bone. (*C*) Selection of *Allosaurus* teeth found at the site. (*D*) The author sitting between some of the bones of the *Camarasaurus*. (*E*) An *Allosaurus* tooth found by the author and Levi Shinkle.

([*A–B, E*] Photographs by the author; [*C*] photograph by Levi Shinkle, courtesy of the Wyoming Dinosaur Center; [*D*] courtesy of Bill Wahl)

FIGURE 4.20. (Overleaf). *The formidable opportunists.*

An *Allosaurus* pack tears apart a rotting juvenile *Camarasaurus* carcass in a Jurassic feeding frenzy. The churned-up ground indicates that many animals have passed by and other carnivores have previously visited the scene.

BOB NICHOLLS 2020

HELL PIG MEAT CACHE

Many animals are known to store excess food in a secure or hidden place and come back to it at a later time, similar to how we store leftovers in a refrigerator. This is known as *food caching* or *hoarding*. Rodents are notorious for this behavior, especially when preparing for the winter months, when food is less plentiful. Some animals cache their food for only a short time, during which they consume it. One of the most famous examples is the leopard, which, after making a big kill, often drags it up a tree, away from other predators, where it is safer to eat. There is some evidence of fossil food caches, namely, pockets full of fossil seeds and nuts. But one surprising, albeit grisly, meat cache was made by a giant "killer *pig* from hell."

"Hell pigs" belong to a group of extinct omnivorous mammals called *entelodonts*. Their nickname is a bit of a misnomer, though, because, despite their pig-like appearance and some shared features, their anatomy pinpoints hippos and whales as their closest relatives.

One of these entelodonts is identified from lots of fossils found in a vast geologic deposit known as the White River Group (White River Formation), exposed in the Big Badlands of South Dakota and neighboring states. Called *Archaeotherium*, at 2 meters in length and about 1.2 meters at its high, humped shoulders, this ancient cow-sized beast somewhat resembled an overgrown warthog and was among the top predators in its ecosystem. It had a massive head, with large, protruding teeth and powerful jaws that could open extremely wide, as much as 109 degrees. Just the sight of one of these was probably enough to intimidate you, especially if you were a small, snack-sized mammal.

The White River deposits date from the latest Eocene to the early part of the Oligocene, about 30–37 million years ago, and have become famous for their rich and exceptional mammal fossils. A wide variety of species have been described, including *Poebrotherium*, a small, sheep-sized humpless camel that superficially looked like a miniature version of a modern llama. Camels originated in North America, and this species is one of the most primitive. Although its fossils are commonly found, an unusual slab of rock measuring 115 by 110 centimeters that contained multiple chomped-up skeletons was discovered in 1998 near the town of Douglas in east-central Wyoming.

The 33-million-year-old kill site contains one complete skeleton, six partial skeletons, and numerous isolated bones from other individuals, all stacked together. A total of 594 bones are exposed, although as many as 700 might be preserved. Bite marks cover the skulls, necks, and thoracic/lumbar vertebrae. The diameter and depth of the puncture marks, along with the spacing and width between punctures, perfectly match the dentition of an *Archaeotherium*—the killer.

Six of the camel carcasses have been sheared in half and do not preserve their rear ends, including the pelvis, back legs, and feet, suggesting that these portions were preferentially consumed. The position of the bite marks suggests that *Archaeotherium* attacked and killed the camels by snapping at the back of their skull and neck before chomping on the meaty hindquarters and tossing the front half into the food cache for later consumption. Similar bite marks are commonly observed on other mammal skeletons from the White River fauna, including on *Archaeotherium* skulls, which tells us that they actively battled each other, perhaps over kills.

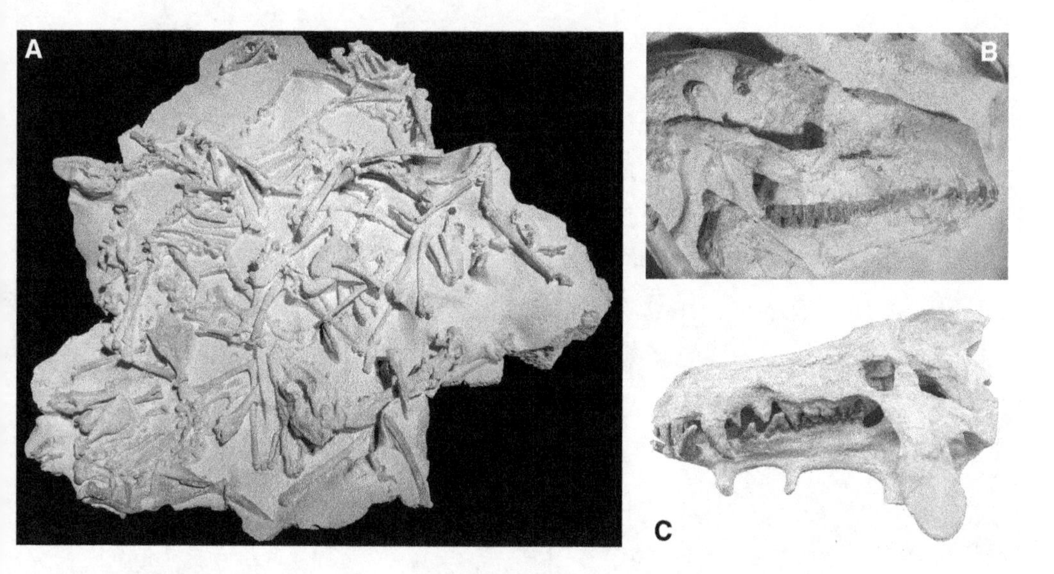

FIGURE 4.21. (*A*) The *Poebrotherium* camel meat cache containing multiple chomped-up skeletons. (*B*) Close-up of one of the bitten and damaged camel skulls. (*C*) Skull of an *Archaeotherium* from Wyoming.

(Photographs by Levi Shinkle, courtesy of the Wyoming Dinosaur Center)

BOB NICHOLLS 2020

This style of caching food is common behavior among modern predatory animals, referred to as *larder hoarding*. It can be advantageous because the animal needs only to remember a single place where it hid the food. On the flip side, it might need to guard the stash and be prepared to defend it. Sometimes predators will kill animals even when ample food is available. This behavior is called *surplus killing* and might have been undertaken by *Archaeotherium* when its camel prey was in abundance. There also remains the possibility that the stash of food was prepared by a parent for its young or maybe even by a pack of hell pigs, although it is impossible to say.

Considering how well preserved the piled *Poebrotherium* skeletons are, in light of their having being caught, slaughtered, and partially eaten, it's possible to suggest that they were buried rapidly, within a short time after being placed in the cache. *Archaeotherium* initially might have covered the kill to conceal it from other predators, but the prey was subsequently buried so deep by sediments that the predator could not find or reach it. Although this carved-up camel meat cache captures a rather gruesome act of behavior, fossils like this offer something more than a simple "who ate whom" relationship.

FIGURE 4.22. (Opposite). *The carcass collector.*

An intimidating *Archaeotherium* hell pig stands above a pile of rotting *Poebrotherium* body parts and is swallowing the back end of one whole.

PREHISTORIC DOLLS: FOSSIL FOOD CHAINS WITH A TWIST

Have you ever seen those Russian matryoshka dolls, the traditional wooden figures that separate in the middle only to reveal a slightly smaller one nestled inside, which then pops open? This process goes on until, finally, a mini figure is found inside. Sure, the dolls are not eating each other, but they serve as a fun analogy for animal food chains. Think of a top predator, such as a killer whale—in this case, the outer shell of the doll—which eats a seal, which has eaten a squid that ate a fish that ate krill that fed on phytoplankton (the start of the food chain, aka, the minute figure). Inevitably, interpreting prehistoric food chains can be extremely complicated and ultimately speculative, except in a few incredible cases in which we are presented with the most extraordinary evidence.

The first multi-fossil food chain to be recognized was brought to light in 2007 and happens to be the most ancient discovered so far. Collected from 295-million-year-old Permian rocks in the town of Lebach in southwest Germany, the fossil comprises the incomplete skeleton of a freshwater shark called *Triodus sessilis*, a member of an extinct family known as xenacanths. Because sharks are cartilaginous, only their teeth are usually recorded as fossils, although this specimen was found inside a mineralized (siderite) concretion that aided the fine preservation of its jaws, scales, and other parts of the skeleton, as well as its teeth. Most surprisingly, the larval remains of two species of tiny temnospondyl amphibians (*Archegosaurus decheni* and *Glanochthon latirostre*) were contained in the gut of this roughly 50-centimeter shark. Unbelievably, the *Glanochthon* also had its last meal preserved, consisting of the partially digested bones of a juvenile spiny shark-like fish called *Acanthodes bronni*.

These animals lived in a deep and immense, 80-kilometer-long ancient lake known as Lake Humberg, where a wide variety of species thrived. The predatory *Triodus* was a common inhabitant of the lake, but it was small fry compared with other, much larger xenacanths and adult amphibians, which would have preyed on it. As it was unable to compete with these apex hunters, *Triodus* likely carved its own niche as an ambush predator, avoiding large adults while hunting their young in shallower parts of the lake.

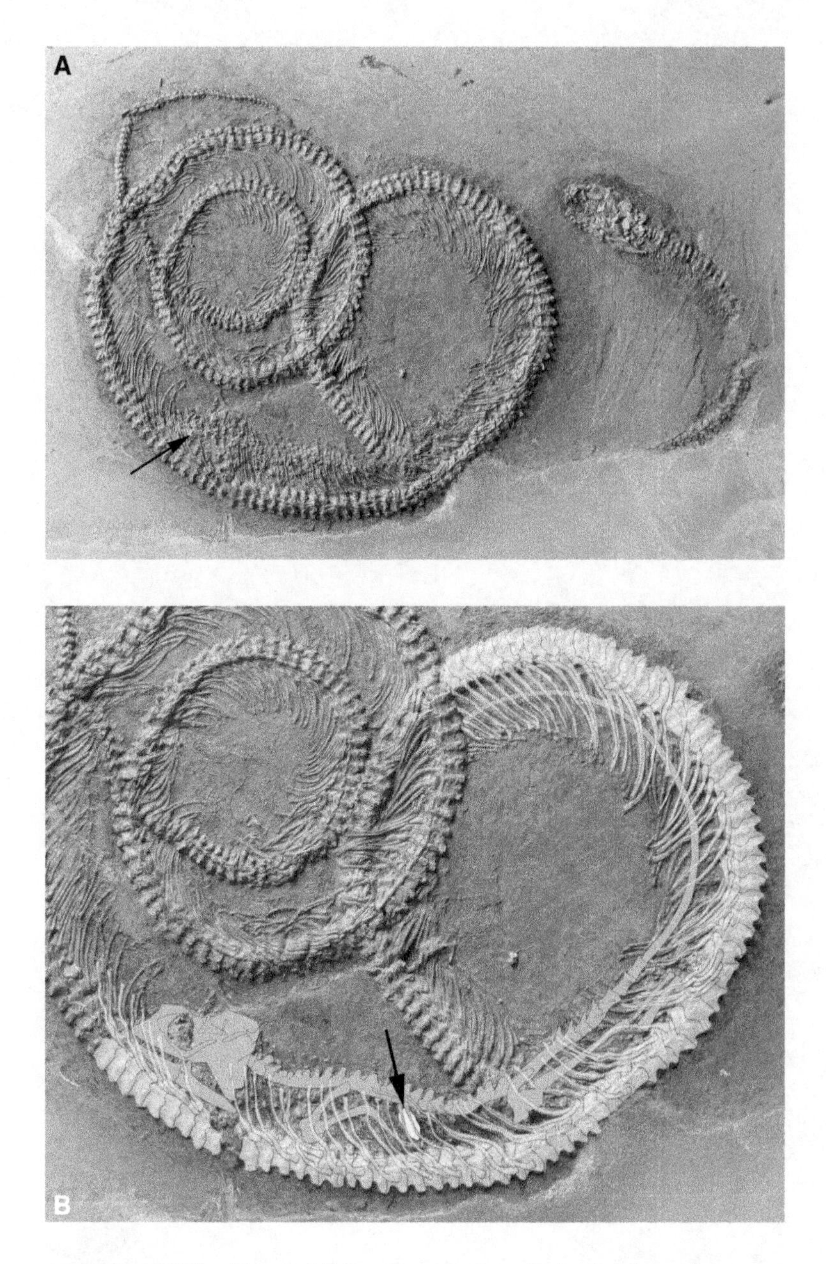

FIGURE 4.23. *(A)* The "fossil doll," an *Eoconstrictor fischeri* snake with a complete basilisk lizard (*Geiseltaliellus maarius*) dinner inside and an unidentified beetle meal preserved inside the lizard. Arrow points to the head of the lizard. *(B)* Overlying illustration showing the position and outline of the lizard and the beetle. Arrow points to beetle.

([*A*] Courtesy of SGN, photo by Sven Tränkner; [*B*] drawing courtesy of Krister Smith/Anika Vogel/Juliane Eberhardt)

The larval amphibians were also a very common species in Lake Humberg and presumably adopted a similar feeding strategy, although in their case, the hunter became the hunted. How both amphibians are oriented in the gut suggests that *Triodus* attacked from behind and swallowed its prey tail first. Their excellent preservation, however, indicates that *Triodus* must have died shortly after it had consumed them, whereas the fragmentary, partly digested fish inside *Glanochthon* hints that it was eaten well before the amphibian itself was caught. Incidentally, no modern shark species are known to feed on amphibians, highlighting the unusual behavior observed in this ancient shark.

In 2009, a similarly preserved vertebrate fossil food chain, only the second to be identified up to that time, was collected from the world-famous Messel Pit quarry in west-central Germany. In contrast to the aquatic interaction just described, this 48-million-year-old Messel meal records direct evidence of a predator–prey relationship in a terrestrial setting. This spectacular fossil includes a complete coiled snake belonging to the species *Eoconstrictor fischeri* (then called *Palaeopython*), a type of early constrictor related to modern boas, that contains the skeleton of an arboreal basilisk lizard (*Geiseltaliellus maarius*), which in turn contains its last insect meal in the abdominal region.

Lizards are the favored prey of most juvenile boa snakes today, especially arboreal species. This *Eoconstrictor* is a juvenile with a total body length of 1 meter, about half the size of known adult specimens from Messel. It is five times larger than its lizard prey, which probably had a torso diameter of 17 millimeters and which was swallowed head first; it is contained inside the stomach, positioned 53 centimeters from the snake's mouth. There is a noticeable kink in the lizard's spine near the pelvis, which might be the result of an injury inflicted by the snake. The eaten insect is a type of beetle, and despite its poor preservation, shows original (structural) color, which shimmers blue-green (as is well known for Messel insects). The lizard must

FIGURE 4.24. (Opposite). *The hunters and the hunted.*

A basilisk lizard, *Geiseltaliellus maarius*, readying to pounce on its beetle prey as an *Eoconstrictor fischeri* snake slowly approaches from behind and lines up an attack.

have eaten the insect a short time before being eaten itself. In a fashion similar to the Permian fossil, the lizard is in excellent condition, with no signs of corrosion from stomach acids. Compared with the digestive speeds in living snakes, this suggests that *Eoconstrictor* died within at least forty-eight hours of its final meal.

How the snake ended up in the lake is another question. Was it swimming across the lake, or did it fall from a tree (maybe after having consumed the lizard)? Or was the snake itself preyed on by a bird that accidentally dropped it into the lake as it flew above? We will never know what circumstances led to the snake's demise, but as with all Messel fossils, it must have fallen deep enough into the lake that it entered the deadly toxic waters at the bottom before it was slowly covered by sediment and became fossilized.

The chance of finding a fossil with its last meal that also contains its prey's last meal is incredibly slim, nearly impossible. Timing and preservation are everything in these three-stage trophic interactions, from the moment these animals dined on each other to when they died and became buried. Just to emphasize that point, here we have indisputable evidence that a small-bodied Permian shark ingested two larval amphibians, one of which preyed on a juvenile fish, and that an Eocene snake ate a lizard that caught an insect. These astonishing fossils offer a superior, definitive direct insight into understanding ancient animal food chains millions of years in the past.

5 | Unusual Happenings

Have you heard about the parasitic crustacean that replaces the tongue of its fish host? Yes, you read that correctly. Amazingly, there are many living species of these tiny isopod crustaceans (called *cymothoids*) that, during their juvenile stage, sneak inside the gills of several types of fish (such as snapper and clown fish). As adults, males stay attached to the gills, whereas the female resides inside the mouth. Once in place, she severs the tongue by cutting off its blood supply before firmly attaching herself to the remaining stub as a functional replacement tongue! Despite losing its tongue to this hijacker, the fish continues to live a pretty normal life, and the parasite grows along with its fishy host, making a home and even starting a family (after mating in the mouth). Not strange enough? What about the toad fly, which has found the perfect spot to lay its eggs—in the nostrils of a toad? When they hatch, the larvae feast on the toad, eventually killing it.

These bizarre acts of parasitism epitomize the unusual intricacies of the natural world. Parasites differ considerably in their causes and effects, from merely bringing their host little harm to causing major disease and death. For these reasons, parasitism was not included in the previous chapter, although parasites feed on their hosts in some way or another, either directly or by eating the host's food.

It can be easy to associate parasites, illness, and other ailments with afflictions that affect only humans, but animals and plants get sick, too. This chapter is different from the others, because instead of focusing on a specific type of behavior or facet of the natural world, it covers a broad spectrum of unusual happenings (health being just one of them). Such happenings include those common or rare aspects of an animal's life that we tend not to think about, whether it's routine things like resting and sleeping, drinking and urinating, breaking bones, getting lost and trapped, and being subjected to the extremes of natural phenomena such as tsunamis.

In all honesty, "unusual happenings" is a bit of a misnomer, because not all these examples are unusual in a modern setting (e.g., getting sick, breaking a bone, or taking a nap), but they make for rather unusual fossils. The story they tell goes against the stereotypical picture that we envisage of the prehistoric world, of animals engaged in one of the behaviors covered in preceding chapters. Ordinarily, these animals are depicted in the prime of their lives and in peak physical condition—except, obviously, for those animals being bitten, ripped apart, and eaten.

We can be confident that, just like animals today, *dinosores* and company occasionally got sick, broke bones, took naps, and dealt with deadly mishaps, such as falling down holes or getting stuck in mud. We need not rely on assumptions, because evidence of such happenings is well documented in the fossil record. On a small scale, fossilized parasitic crustaceans have been found preserved inside fish, inferring similar parasite–host associations akin to modern species. On the other extreme, large-scale events are exemplified, such as the presumed dinosaur family trapped in quicksand and the community of animals that were buried and killed by volcanic ash. Both of the large-scale examples would have been included under this chapter were it not for the apparent social and group behaviors recorded in those exceptional circumstances which led to their inclusion in chapter 2.

One area of study that has provided a great wealth of information concerning ancient aches and breaks is *paleopathology*, the study of prehistoric diseases and injuries. When studying fossilized bones, vertebrate paleontologists are often presented with a multitude of evidence for fossil fractures, be it a broken toe, cracked rib, or deformed vertebra. Diagnoses like these might help to reveal how the disease or injury impacted the individual's life: whether it caused minor or major damage and if it had short-term or long-lasting side effects. Sometimes broken bones show signs of healing and suggest that the animal survived the trauma, but others display no signs of recovery and imply that the animal died, perhaps due to a fatal fall or a predator attack.

The broad and diverse nature of fossils included in this chapter reveals unexpected behaviors and situations unlike anything covered previously. So, sit back and prepare to puzzle, ponder, and laugh at some of the most unusual fossils that we *do* have evidence for, from loose liquids to deadly diseases.

PARASITE REX

One of the largest terrestrial predators ever to have lived, at more than 12 meters long and weighing over 8 tons—almost as much as two African elephants—is a full-grown *Tyrannosaurus rex*. This dinosaur had no natural predators, except perhaps another *T. rex*. Without a doubt, Rex is *the* celebrity of the dinosaur world. For most people, it is the ultimate top-of-the-food-chain predator. But as mighty as it might have been, some *T. rex* and kin fell victim to the unlikeliest and tiniest of "predators"—parasites.

Parasites might not spring to mind when we think of fossils, especially as far as *T. rex* killers go, but they have been plaguing animals for millions of years. Living inside (*endoparasitism*) or on (*ectoparasitism*) a host, parasites use it for food and a place to live, sap its energy while they survive and thrive, which can ultimately lead to the host's death.

In 1990, while fossil hunting in the badlands of South Dakota, fossil collector Sue Hendrickson happened on a truly awesome discovery: the world's most complete and arguably most famous *T. rex*. Although we do not know whether this specimen was male or female, it was given the name SUE (which must be in all capitals according to the Rex's Twitter account), after its discoverer. SUE has been the subject of intense study, and one of the most exciting revelations surrounds the death of this enormous predator.

SUE's lower jaw bears a series of abnormal, smooth-edged erosive "holes," which have stumped paleontologists for years. Previously they were identified as bite marks or as possibly symptomatic of a bacterial bone infection, but in fact, neither is the case. These features are also not unique to SUE. As part of a wider study, additional specimens of *T. rex* and other members of the tyrannosaurid family (including *Daspletosaurus* and *Albertosaurus*) were examined, and similar-looking lesions were found in the same area of their jaw bones.

These lesions bear a striking resemblance to those found in the lower jaw of modern birds, commonly pigeons, doves, and chickens but also birds of prey, and result from a common parasitic infectious disease called *trichomonosis*. The parasite effectively eats chunks of the jawbone. This extremely nasty condition causes severe damage and pain around the mouth, throat, and esophagus, making simple things like eating and drinking unpleasant to nearly impossible. Ongoing research focused on SUE suggests that the infection also could have led to some abnormal tooth development which might have caused additional pain in the jaw.

This discovery represents the first time that an avian (bird) transmissible disease has been found in a non-avian theropod dinosaur. Considering the similarities between the lesions in tyrannosaurs and modern birds, the trichomonosis-like disease in *T. rex* and kin suggests that these dinosaurs

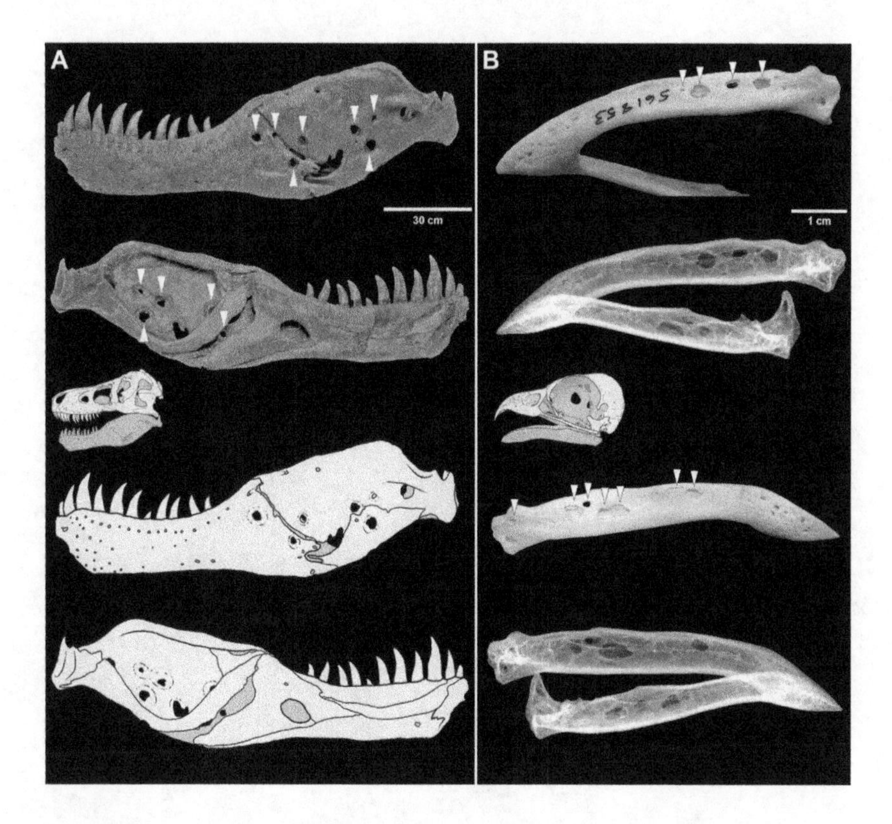

FIGURE 5.1. (*A*) Photographs and illustrations of the left lower jaw of *Tyrannosaurus rex* SUE displaying multiple trichomonosis-type circular lesions, identified by the arrows. (*B*) Photographs and radiographs of the lower jaw of a modern bird (osprey) displaying strikingly similar circular lesions resulting from the trichomonosis parasite, identified by the arrows.

(Images courtesy of Ewan Wolff. [*A*] Photographs by John Weinstein, © Field Museum. All are modified slightly from Wolff, E. D. S., et al. 2009. "Common Avian Infection Plagued the Tyrant Dinosaurs." *PLOS One* 4, e7288)

were susceptible to similar, or the same, unpleasant complications of the disease and had a similar immune response to it. It is even possible that the infection is the result of the same trichomonosis-causing parasite that affects birds today. Once the animal was infected, feeding would have been difficult, and it is highly likely that, as seen in living birds, the mighty tyrannosaurs lost considerable weight before eventually starving to death.

The disease was probably endemic. Although there are several ways in which it could have been transmitted, such as through consumption of infected prey, it seems that face biting was the primary cause of transmission, which fossil evidence suggests was a common behavior among tyrannosaurs associated with competition, territoriality and courtship. Similarities can be seen with the living Tasmanian devil, which is at risk of extinction in the wild due to a contagious oral cancer known as devil facial tumor disease. Once the disease is transmitted through face biting, diseased individuals usually die within six months.

Being infected by such a deadly disease would have severely limited what the animal could do in its day-to-day life. Over time, the parasite would have felled its mighty host. It seems almost like some strange fairy tale that these tyrannosaurs, which surely terrorized countless animals in their lifetime, were themselves plagued by a predator, yet one so tiny that *T. rex* could not even see it.

FIGURE 5.2. (Opposite). *Trichomonosis rex.*

Looking worse for wear, SUE, the world's most complete *Tyrannosaurus rex*, was severely infected by a trichomonosis-like deadly disease.

MASS MAMMAL STRANDINGS: AN EPIC ANCIENT GRAVEYARD

Marine mammal strandings are common phenomena and occur on a global scale. This peculiar form of behavior, also known as *beaching*, is a sad event that can involve many hundreds of individuals. There is no single explanation as to why they are found in such situations either alive, injured, or dead. Aside from human causes, like the use of military sonar and chemical pollution, in many cases, these strandings are due to natural causes such as fatal errors in navigation, disease, extreme weather, and deadly toxic waters. Prehistoric marine mammals were also subjected to strandings, and the most extraordinary discovery was made by chance in Chile's Atacama Desert.

In 2010, construction workers in northern Chile, close to the port of Caldera, uncovered a large fossil site during the expansion of a road along the Pan-American Highway, which is part of an extensive series of roads that stretches from Alaska to Argentina. Some of the fossilized bones hinted that this site—called Cerro Ballena, Spanish for "whale hill"—would be special, so it was temporarily opened up as a 20 by 250 meter quarry. A team of North and South American paleontologists led by marine mammal expert Nick Pyenson of the Smithsonian National Museum of Natural History in Washington, D.C., were given just two weeks' access to inspect, excavate, and study the remains as quickly but carefully as possible. It was very much a salvage operation.

A huge number of fossils were found, including more than forty complete and partial marine mammal skeletons, aquatic sloths, billfishes, and isolated shark teeth. All of these fossils come from 6 to 9-million-year-old late Miocene marine sediments of the Bahía Inglesa Formation, which is well known in the Atacama region.

The quarry site really only scratched the surface of what lies beneath. Unfortunately, much of the site no longer exists, as it is was paved over and now sits under the new road. Given the rich extent of the material found in a relatively small area, the original quarry exposure represented only a fraction of what surely is a far more extensive fossil site; geologic maps indicate that it had an extent of approximately 2 square kilometers. Thus, it is highly likely that hundreds more skeletons are still buried there.

The diversity of marine mammals at Cerro Ballena is particularly impressive. The most abundant is a large baleen (rorqual) whale, a member of the same large group that contains the living blue whale. It is identified from at least thirty-one individuals of different ages, including young calves to mature adults (up to 11 meters long), which probably all belong to the same species, although studies of the fossils are still ongoing, and the species has yet to be identified. Others include extinct types of at least two different seals, a single sperm whale, and a walrus-like toothed whale (called *Odobenocetops*). The cetaceans are represented by complete, articulated skeletons, and, notably, many of the baleen whales were found facing the same direction in a mostly belly-up position that, in conjunction with their abundance, completeness, and fine preservation, strongly suggests that they washed up either dying or dead, corresponding to a modern stranding event. A live-stranded individual is typically oriented upright because of its blowhole.

This prehistoric baleen whale was obviously social, as are many living whales. However, although mass strandings of modern toothed whales are common, mass strandings of modern baleen whales are comparatively rare. In one example documented between November 1987 and January 1988, over a period of five weeks, a total of fourteen humpback whales were stranded along the coastline of Cape Cod, Massachusetts. These individuals differed in sex and age (including a single calf), and no signs of injury were observed. Their last meal (Atlantic cod), however, remained inside their stomachs and was revealed to contain high levels of algal toxins, which fatally poisoned the whales. Direct observations showed that they died quickly at sea.

The marine mammal fossils were found at four distinct bone-bearing levels identified within an 8-meter section of the formation, each reflecting individual mass stranding events separated over a period of several thousand years. At each level, the mammals are preserved in close proximity, with some in direct contact, suggesting that these individuals were all subjected to the same sudden cause of death at sea. Subsequently, during harsh storms or spring tides, they were cast onto a tidal mudflat and buried. Today, harmful algal blooms, or HABs (sometimes called "red tides"), which account for most modern strandings, are the only explanation for

FIGURE 5.3. (*A*) Paleontologists excavate several baleen whale skeletons at the Cerro Ballena mass graveyard adjacent to the Pan-American Highway. (*B*) A closer view of three skeletons being excavated, with overlapping adult and juvenile specimens.

([*A*] Photo by Adam Metallo, Smithsonian Institution; [*B*] image modified slightly from Pyenson, N. D., et al. 2014. "Repeated Mass Strandings of Miocene Marine Mammals from Atacama Region of Chile Point to Sudden Death at Sea." *Proceedings of the Royal Society B* 281: 20133316) such repeated accumulations of multiple individuals and different species. Evidence for ancient algae is also contained in the rocks that entombed the fossilized bones.

HABs occur when certain algae grow to immense, uncontrollable sizes while producing powerful, deadly toxins. The ecosystem's health can be so greatly affected that animals are unable to survive. In marine mammals, the poisonous toxins result in organ failure and, ultimately, death. Entire herds can be (and are) wiped out. For example, HABs, combined with a building El Niño, were to blame for the largest reported mass mortality of baleen whales (a staggering 343 individuals, primarily sei whales, recorded in 2015 in southern Chile).

The abundance of exceptionally preserved, associated skeletons found at Cerro Ballena makes it one of the richest fossil marine mammal localities in the world. Fossil evidence and comparisons with modern analogues point to this graveyard as representing four recurring mass death assemblages due to HABs. It is assumed that the prehistoric species were poisoned by toxic algae, through ingesting contaminated prey and/or inhalation, which led to their rapid deterioration at sea. The dying or dead bodies then drifted to the coastline, where they were deposited and buried. This discovery illustrates that prehistoric marine mammals were prone to mass stranding events and that this species of baleen whale was also social. Unfortunately for them, as today, their strong, tight-knit social bonds might well have led to their downfall, for they not only lived and traveled together but, in this case, also ingested deadly toxic algae and died together.

FIGURE 5.4. (Overleaf). *Victims of the bloom.*

A mass stranding of numerous baleen whales, a few sperm whales, seals, and multiple species of fish, caused by lethal toxic algae from a harmful algal bloom.

SOUNDLY SLEEPING DRAGON

Sleep and rest are essential for the brain and body to recharge, get stronger, and function normally. Some animals sleep for most of the day, others rest periodically, and certain birds and marine mammals switch off half of their brain while the other side remains awake. Birds are even known to rest or sleep during flight. These behaviors differ considerably across the animal kingdom, and this complexity can make it difficult for scientists to research and understand sleep behaviors. This is why the breathtaking discovery of a small, birdlike dinosaur preserved in an unmistakable sleeping or resting position typical of many modern birds is such an exceptional find—a real sleeping beauty.

Mei long, meaning "soundly sleeping dragon" in Chinese, is a 53-centimeter-long, chicken-sized carnivorous theropod that was unveiled to the world in 2004. *Mei* is a member of the birdlike troodontid family, close relatives of *Velociraptor* and kin, and its practically complete skeleton was discovered in the fossil-rich, 125-million-year-old Early Cretaceous rocks of Lujiatun in western Liaoning, China.

Mei is preserved in a remarkably lifelike pose, curled up with its tail wrapped around the body and under the neck. Its body rests on long, folded legs; the forearms are folded against its sides just like a bird; and its small head is positioned to the left, tucked neatly between its left forearm (at the elbow) and the body. This is a pose that matches the characteristic sleeping or resting posture found in living birds. This tucked-in head pose also helps to conserve heat in birds, and that might well have been the case for *Mei*, implying that this behavior first evolved in non-bird dinosaurs. Only the bones are present, and they show that *Mei* was a young adult, but its anatomy, coupled with comparisons with other family members, confirms that it would have had feathers.

The find represented a world-first in dinosaur research. And yet, this fossil is not unique. Since its discovery, a second similarly preserved and complete *Mei* has been found, lying in an identical position but with its head tucked to the right. At least two other specimens exist that probably represent this species, but, they have yet to be formally identified and studied. Other birdlike dinosaurs have also been discovered in similar postures, suggesting that the pose was not simply random but probably a common sleeping and/or resting position typical of these animals.

Curiously, some birds enter rapid eye movement (REM) sleep, the same type of sleep in which most dreams occur in us humans. We are unable to confirm if birds dream, but research infers that they do; in fact, one study found that zebra finches practice songs by singing in their sleep. If this is the case, although it is rather poignant, could it be that these birdlike dinosaurs were dreaming when they perished peacefully?

It is impossible to say whether *Mei* was actually sleeping or just taking a rest, let alone dreaming. Whether *Mei* was sleeping or resting, you probably wonder how it was preserved in such a state. Initial studies of the site (in Lujiatun) where it was found suggested that it, along with the dinosaurs and other animals that shared its environment, were killed and buried by hot, airborne volcanic debris and ash in a single mass death event. The site has

FIGURE 5.5. The first skeleton of the "sleeping dragon," *Mei long*, preserved in a typical birdlike sleeping or resting pose.

(Courtesy of Mick Ellison, American Museum of Natural History)

been aptly labeled the "Chinese Pompeii." However, recent findings have shown that the fine preservation of so many spectacularly complete and three-dimensionally preserved skeletons represents multiple events rather than a single mass death. It is thought to be the result of a combination of flood events mixed with high loads of volcanic debris resulting from ashfalls, lahars (volcanic mudflows), and pyroclastic flows, which smothered and rapidly buried the animals. It is possible that they had already died by asphyxiation from toxic volcanic gases or could have been buried alive while sleeping or resting on the ground. Given the lack of disturbance to some specimens, like *Mei* found in its original lifelike pose, it has been suggested that the animals might have been rapidly entombed inside a collapsed underground burrow (or equivalent) that was not preserved. Although plausible, without any direct evidence, this remains speculative.

The thought of finding a sleeping or resting dinosaur seems outrageous, if not impossible, so finding multiple examples is extraordinarily rare. We can assume with confidence that dinosaurs routinely slept or rested as part of their daily lives, like animals do today, including living dinosaurs. The birdlike posture of both specimens forms yet another behavioral link between birds and their very ancient relatives. These exceptional "sleeping dragons" not only looked like birds but provide confirmation they also slept and/or rested like birds, only for paleontologists to wake them up millions of years later.

FIGURE 5.6. (Opposite). *The dragon needs a blanket of ash.*

Ash begins to fall like snow around a curled-up, soundly sleeping *Mei long*.

SNAP THAT! A DRAMATIC JURASSIC CROC

Vertebrates commonly sustain broken bones. This can be agonizingly painful and force the sufferer into an extended period of rest, and, depending on the type of break and its severity, could leave them incapacitated and with a heightened risk of death. Yet, given time, bones have the remarkable ability to heal, allowing for an individual's potential recovery.

With 206 bones in the adult human body, it is not surprising that people break at least one of them during their lifetime. Various male mammals, such as cats, dogs, bears, bats, rodents, and seals, even possess a bone in their penis, called a *baculum*. (The female counterpart is called the *baubellum*, found in the clitoris of some species.) And yes, that bone is known to snap occasionally under strain during sex. It can heal, although sometimes crookedly; even snapped fossil bacula have been found. Considering the enormous number of fossils that show some form of bone damage, thus reflecting life events, it is difficult to choose just one example. However, a gnarly fossil "crocodile" with a shocking deformity was collected from a Jurassic quarry in the town of Dotternhausen, near the famous fossil sites of Holzmaden, Germany.

The fossil is that of a practically complete *Pelagosaurus typus*, a member of an extinct group of crocodile relatives called *thalattosuchians*, commonly referred to as marine crocodiles. Most are characterized by an exceedingly long and narrow snout, outwardly resembling the modern-day gharial (or gavial) crocodile, which inhabits rivers in India and Nepal. Measuring just a couple of meters long, *Pelagosaurus* is a dwarf species that spent most of its time in the warm, shallow waters, presumably coming ashore to lay its eggs and rest.

Looking at the skeleton, your eyes are immediately drawn to the beautifully preserved skull, where it becomes obvious something is very wrong. Halfway along the delicate and toothy snout, the lower jaw is snapped and pointed at an almost 90-degree angle relative to the skull. This is not the result of a postmortem fracture, nor is it damage from the excavation process. Rather, the presence of new bone in the form of a massive bony callus at the base of the fracture shows that this croc initially survived a major trauma. It is evidence that, immediately following the injury, a blood clot

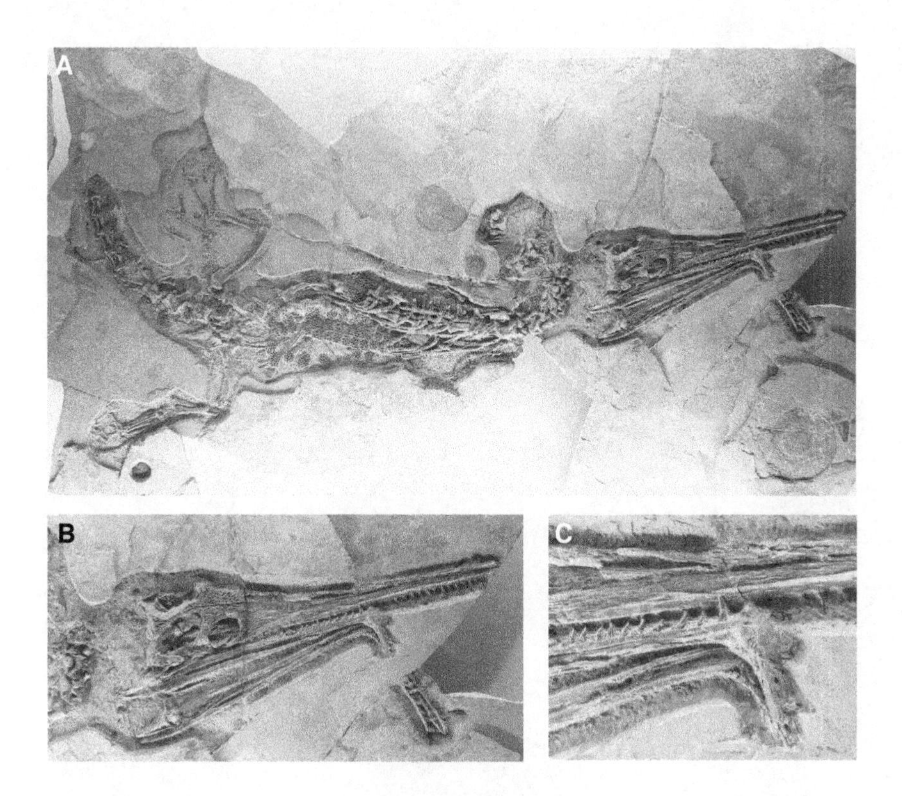

FIGURE 5.7. (*A*) Skeleton of a *Pelagosaurus typus* with a snapped jaw. (*B*) Close-up of the skull revealing the almost 90-degree broken jaw. (*C*) Closer view of the large bony callus.

([*A*] Courtesy of Sven Sachs; [*B–C*] photographs by the author)

FIGURE 5.8. Living crocodile with a large portion of its upper jaw missing.

(Photograph by Sailu Palaninathan, courtesy of the director of Arignar Anna Zoological Park, India.)

(hematoma) formed around the broken bone to protect it and initiate the healing process. As healing continued, the damaged portions of the mandible were united by the formation of a soft callus, which eventually hardened into a solid bony mass.

Unfortunately, because the two halves of the broken mandible were not realigned correctly during the formation of the callus, the bone healed in an abnormal position; this is referred to as a *malunion fracture*. It thus became a major hindrance for the croc, which had to do everything in this fixed position, be it swimming, hunting, or walking. Swimming would have been especially awkward, perhaps painful, due to the amount of drag force placed on the jaw. Equally, it would have been particularly cumbersome on land, as the croc would have needed to hold its head off the ground or sideways at all times to prevent the jaw from being snagged.

Crocodilians are extremely hardy animals, known to survive major injuries. Among modern species, fights are often particularly violent and might lead to death. The combination of biting with crushing force, death-rolling their rival, and using their head as an effective bashing weapon can lead to serious, bone-cracking damage whereby jaws are often fractured and sometimes torn completely off. Still, some are able to live normal lives even with sections of their upper or lower jaws missing. In one example, the late Australian wildlife expert and TV host Steve Irwin published a brief scientific paper describing a saltwater crocodile (nicknamed "Nobby") that he captured with a significant portion of its lower jaw missing; part of its tongue had also been amputated. The croc had been known to locals for at least eighteen years and had apparently survived by exploiting a nearby cattle station rubbish dump, where it scavenged the rotting carcasses of various animals. Following inspection, the individual was released back into the wild. Gharials also have been observed with severed jaws, and even whales have been spotted with similarly broken mandibles. It seems most likely

FIGURE 5.9. (Opposite). *Against the odds.*

A disfigured *Pelagosaurus* crosses the beach holding its head high to keep from snagging its awkwardly fractured jaw on the ground.

that the fracture in *Pelagosaurus* was sustained during a fight with another of its kind.

Despite the horrific nature of the injury, the bony callus confirms that this individual survived for at least a time with the downturned jaw, perhaps weeks, months, or longer. The unfavorable position would have severely restricted its ability to feed, especially considering that *Pelagosaurus* is postulated to have hunted like the gharial, slowly stalking its prey before attacking with a swift sweep of the head, snatching fast-moving food such as fish.

Considering that the injury rendered this individual a sitting duck, it is surprising that it did not become prey for any of the various larger marine reptiles that lived alongside it, including ichthyosaurs (some giants were up to 12 meters), plesiosaurs, and other marine crocs. This competition from and avoidance of predators, combined with the major injury, probably led to the animal's starving to death. Rather counterintuitively, the healing of a broken bone was a death sentence for this 180-million-year-old marine crocodile, which might have survived longer had the fractured jaw fallen off.

THE DRAMA OF A DROUGHT?

Mass fossil accumulations can offer numerous insights about ancient communities. Although understanding those communities is significant, working out how and why mass accumulations formed is equally important. Were they the result of a catastrophic event, such as a volcanic eruption that caused the animals to die simultaneously, or was it something less momentous, like individual carcasses being washed ashore and buried together over time? Examination of the rocks containing the fossils can help to reveal the cause behind their creation, but deciphering exactly what happened and offering a plausible explanation for their preservation can be hard to do.

In 1939, legendary American paleontologist Alfred Sherwood Romer, best known for his major contributions to the study of vertebrate evolution, described a spectacular accumulation of fossil amphibians. At the time, they were attributed to the species *Buettneria perfecta*, which later became known as *Koskinonodon perfectum* and is today called *Anaschisma browni*. (The naming game can sometimes be complicated.) Up to 3 meters long, *Anaschisma* looked like a supersized salamander, with a large, flat head. It is part of an extinct family known as metoposaurs, which filled the role of crocodile-like predators in ancient lakes and rivers during the Late Triassic.

The *Anaschisma* assemblage was found by Robert V. Witter and his wife in 1936 as they were exploring the 230-million-year-old Triassic rocks just south of Lamy in Santa Fe County, New Mexico, as part of a fossil expedition for Harvard's Museum of Comparative Zoology. Initially, the Witters found fragmentary bones that had trickled down the side of a small hill and traced them to an extensive and densely packed bone bed crammed full of amphibians, which was overlain by a mass of sandstone. Two years later, with the aid of picks, shovels, and a few sticks of dynamite, excavation revealed that the bone layer extended 15 meters or more along the exposure, but it measured just 10 centimeters in thickness.

As many as one hundred adult individuals might be preserved, represented by at least sixty good skulls (about 60 centimeters long), along with parts of the skeletons, all piled atop and overlapping one another in a jumbled mass. Romer assumed that the locality would have been much larger were it not for erosion and suggested that hundreds, if not thousands, of these large amphibians might have been buried together. Until the discovery, very few of these amphibian fossils were known from North America, and the site became famous in paleontology circles as the Lamy Amphibian Quarry.

Romer proposed that the amphibians died in a pond that dried up during an intense drought, which had forced the individuals to congregate in the last remaining pools of water. In these crowded conditions, the few remaining survivors struggled for space among the corpses and eventually died of either starvation or the complete drying of the pond.

The pond-receding, drought-induced idea paints a rather vivid picture that seems plausible but lacked any real evidence. Only in the 1980s and '90s was this classic interpretation questioned, and in 2007 (sixty years after the last dig, in 1947), an excavation was undertaken by members of the New Mexico Museum of Natural History, who uncovered new data that challenged Romer's hypothesis further. Those studies found no evidence of a drought or a pond deposit at the site and suggested that the mass mortality was instead due to an unknown cause. It is possible, however, that a drought might have been the initial reason that this group accumulated (and died), but it was not thought to be the reason for their eventual burial and preservation. Regardless, the bone bed does represent a catastrophic mass death event. The disarticulated and disassociated nature of the amphibian skeletons suggests that the carcasses were already fully (or mostly) decayed prior to being rapidly transported, then deposited, and eventually buried on a nearby soil-forming floodplain.

We have no way to be certain what form of gregarious behavior this enormous group of amphibians were engaged in prior to their mass death. However, many living amphibians lay their eggs communally and mate in large communal groups. Given the adult-only association at Lamy, it seems that this site might represent a breeding population. Lamy is not a lone occurrence. Several other mass metoposaur death assemblages are known from the Late Triassic, recorded at other sites, including in the western United States (such as the Rotten Hill Bone Bed in Texas, which includes numerous *Anaschisma*), Morocco, Poland, and Portugal. Each apparently contains only adult amphibians.

The Moroccan site is situated in Argana and differs from the rest, in that seventy large and mostly fully articulated and complete individuals (of the species *Dutuitosaurus ouazzoui*) were present in the bone bed. Ironically, as Romer had proposed for Lamy, this death assemblage was likely to have been caused by a drought that led to the drying of a pond, as further evidenced by the identification of ancient mud cracks that surrounded the quarry where the fossils were found. Mud cracks would be expected for a body of water that was drying up under drought conditions. In this case, all of the larger adults were positioned in the center of the deposit (in the deepest/last part of the drying pond), encircled by the smaller individuals, which had been outmuscled and pushed aside.

FIGURE 5.10. Part of the famous Lamy Amphibian Quarry specimen containing numerous skulls and bones of the large amphibian, *Anaschisma browni*.

(Courtesy of Spencer Lucas)

2020

Bone beds that principally contain a single species are especially interesting because they hint at some form of gregarious group behavior. These rare amphibian assemblages show that metoposaurs were social, at least some of the time, aggregating in large groups and regularly succumbing to mass death events. Whatever the reason for their formation, each individual died, was buried, and was subsequently preserved in the same way, resulting in some of the most dramatic fossils ever found.

FIGURE 5.11. (Opposite). *Casualties of the drought.*

Countless *Dutuitosaurus ouazzoui* are crammed into a drying pond, piled on top of one another and trying to keep wet in the remaining muddy water. Some of those on the periphery have already died in the parched conditions.

EATEN FROM THE INSIDE OUT

Wasps have been flying around and pollinating plants since as far back as the Jurassic and probably pestered our early mammal ancestors, too. Just say the word "wasp," and some people will freak out in fear of being stung. In truth, of the more than 100,000 living species most do not sting, act as valuable pollinators, and keep pest populations under control.

One hugely diverse group of wasps are the parasitoids that have evolved a specialized, albeit rather creepy, behavior that makes stinging seem a bit less frightening. They find a suitable arthropod host for their young and lay their eggs onto or inside the victim. The unwilling host might be at any stage of development (from eggs to adults) and remains alive long enough so that the developing young, who live as parasites and feast on the host from the inside out, eventually break free and kill it. In one of the strangest examples, the larva of certain species takes control of their spider host's brain in a zombie-like fashion and force it to spin an unusual web, which acts as a protective "cocoon" so the larva can safely transform into its adult stage. The young wasp then eats the spider.

Evidence for these types of parasitoid behaviors has been found in a few very rare fossils, mainly contained in amber. But the most impressive find was collected from a phosphorite mine in the Quercy region of south-central France. Here, a substantial collection of three-dimensional fossil fly pupae, preserved in their cocoon-like state, were found in the late 1800s and early 1900s and are between 34 and 40 million years old, from the Eocene.

Measuring just 3 or 4 millimeters in length, these tiny fossils resembling rice grains were formally described in the 1940s, and one was recognized as possibly hosting a parasitoid wasp, but they remained unstudied. In 2018, more than one hundred years after the fossils' discovery, paleontologists decided to reexamine the fly pupae with the aid of high-tech equipment, which continues to play a prominent and highly beneficial role in bringing fossils to life.

Powerful synchrotron X-ray imaging was used to peer inside 1,510 fossil pupae, many of which remained covered by hardened skin. This scanning process is nondestructive and allows for the contents to be fully visualized in three dimensions. That sounds like a lot of specimens, but considering their small size, it took just four days to scan them all. Amazingly, 55 of the fossil fly pupae were found to contain parasitoid wasps, and 52 of these wasps were identified as adults.

Many of the wasps are extremely well preserved, right down to the presence of tiny hairs (setae) across their bodies. In some examples, the wasps were fully developed, even showing unfolded wings, indicating that they had hatched into their adult form and were ready to emerge from their hosts, perhaps as a synchronous release, all emerging at the same time. Both males and females could also be identified. A few of the pupae contained fragments of legs and bristles belonging to the developing fly hosts, the bits the wasps did not eat.

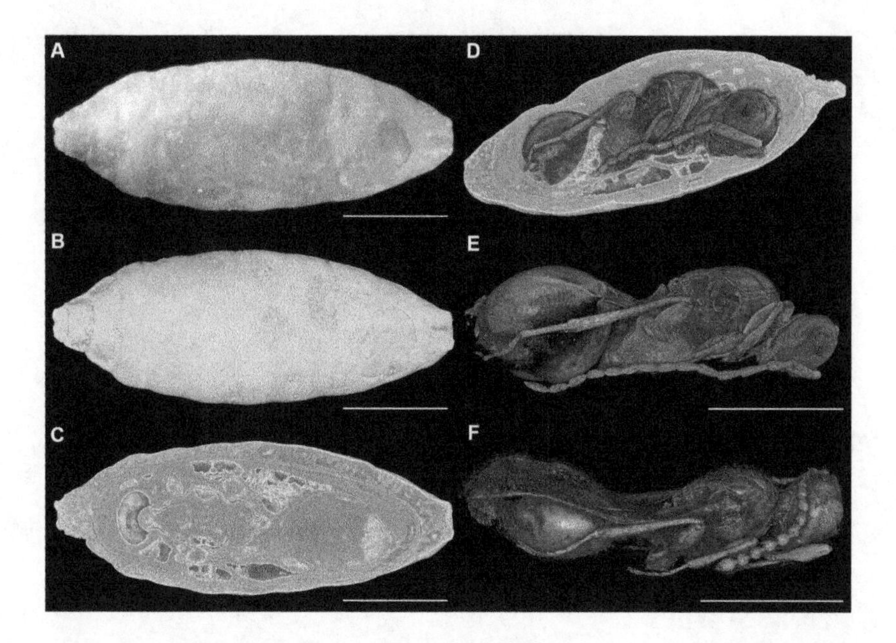

FIGURE 5.12. (*A–C*) A fly host pupa. (*D–E*) The same pupa showing exceptional details of a well-developed male *Xenomorphia resurrecta* parasitoid wasp inside its host, preserved with folded wings. (*F*) An *X. resurrecta* female with unfolded wings. Scale bars measure 1 millimeter.

(Courtesy of Thomas Vandekamp)

BOB NICHOLLS 2020

As a bonus of the discovery, the parasitoids were found to represent four previously unknown types. Befitting their bursting-out behavior, two of the new species were named *Xenomorphia resurrecta* and *Xenomorphia handschini*, after the chest-bursting xenomorph aliens in the *Alien* movie franchise. Clearly, the researchers had a sense of humor.

It is intriguing to ponder the situation in which these pupae were preserved. Just as occurs today, the stench of a rotting carcass probably attracted flies, which laid their eggs on it. The eggs hatched into maggots, which ate their fill of flesh before entering their final preadult form as pupae. During this stage, female parasitoid wasps pierced the squishy pupae with their needle-like ovipositors and injected a single egg. The pupae were each then feasted on by the developing wasp larvae, which grew into adults and spread their wings in anticipation of release. At that stage, just prior to emerging from their fly hosts, they were submerged into phosphate-rich water, which led to their fossilization. When the ancient fly pupae were discovered, the collectors likely never could have dreamt that these apparently common fossils would in the future yield such unusual findings.

FIGURE 5.13. (Opposite). *Condemning the host.*

An adult female *Xenomorphia resurrecta* using its needle-like ovipositor to inject an egg inside a developing fly pupa.

DINOSAUR TUMORS

The human body is a wonderfully complex machine made up of more than 30 trillion cells. When new cells replace the old ones, they might grow and multiply in an abnormal, uncontrollable way and form a tumor. For many, the word "tumor" is synonymous with cancer, but not all tumors are created equal. A tumor might be benign, meaning that it is noncancerous and generally not harmful, or malignant—one that is cancerous and potentially fatal. Cancer has affected us all in one way or another and must be among the most hated six-letter words in the world. It is easy to associate such deadly diseases with humans, but modern animals also develop tumors and die of cancer, and so did dinosaurs.

Various types of bone tumors affect humans and animals, leaving their own unique lesions. Comparing the same features with dinosaur bones, paleontologists are able to detect and diagnose specific types of tumors. This is extremely rare, and only a handful have been recognized.

This rarity is exemplified by the findings of a team led by Bruce Rothschild, a professor of medicine and a vertebrate paleontologist. He is the go-to person when it comes to dinosaur paleopathology, as he spent decades studying and writing about disease in fossils. In 2003, just a few years after the first dinosaur tumors were diagnosed by Rothschild and company, as his team searched for further evidence, they X-rayed more than 10,000 dinosaur vertebrae belonging to more than 700 individuals. A wide variety of dinosaurs from different periods were selected, with members of every major group included, such as *Stegosaurus*, *Diplodocus*, and *Tyrannosaurus*. However, only 29 individuals were found to contain tumors, and, curiously, all samples were tail vertebrae from the same group of dinosaurs, hadrosaurs. (You might be more familiar with their ill-suited nickname, "duck-billed" dinosaurs. Think Ducky from the *Land Before Time* series.) These were a common group of herbivores that lived in large herds and flourished during the Late Cretaceous, some 66–85 million years ago.

Most of the tumors were associated with the school bus–sized hadrosaur *Edmontosaurus*, including one that was identified as metastatic cancer. Known also as secondary cancer, this type spreads to the bones after developing in another part of the body. Thus, this *Edmontosaurus* was affected by a primary cancer of unknown origin and was dying or perhaps died as a result of the advanced cancer.

Besides hadrosaurs, only four other dinosaurs have been unambiguously identified with tumors. Two of these are represented by isolated bones from indeterminate dinosaurs recorded from the Jurassic of Utah and Colorado; the specimen from Colorado also represents a rare case of metastatic cancer and was the first to be diagnosed in a dinosaur (by Rothschild and his team). An interesting example was identified in a tail vertebra from a giant titanosaur sauropod collected from the Cretaceous of Brazil. It was found to contain two different types of benign tumors (an osteoma and a hemangioma). Most recently, in 2020, an aggressive cancerous bone tumor (an osteosarcoma) was diagnosed in the fibula of a ceratopsian dinosaur called *Centrosaurus*, collected from the Cretaceous of Canada. Other possible tumors have been reported in the femur of the titanosaur *Bonitasaura*, collected from the Cretaceous of Argentina, and the femur of the stegosaur *Gigantspinosaurus* from the Jurassic of China.

These tumors might have caused some discomfort and perhaps even substantial pain depending on their size and position, although the advanced nature of the cancer in *Centrosaurus* would have had crippling side effects. One tumor that would have been especially noticeable and probably impacted the individual in life was recognized in a specimen of the primitive hadrosaur, *Telmatosaurus transsylvanicus*. The fossil consists of a well-preserved pair of lower jaws with teeth, collected from along the banks of the Sibişel River in the Haţeg Basin region of Transylvania, Romania (hence the name).

This dinosaur had a facial deformity, the result of a very rare benign tumor called an *ameloblastoma*; the first to be reported in the fossil record. In humans, this usually slow-growing but aggressive tumor is located in the jaw (commonly the lower jaw) near the molars and wisdom teeth and develops from cells that form tooth enamel. In some cases, it can cause severe pain in the jaw, displace teeth, and reach such great size that the shape of the face might be changed dramatically. In *Telmatosaurus*, the tumor is located in the left lower jaw, identified by the presence of a prominent bony bulge with a unique soap-bubble-like internal structure, which is typical of ameloblastoma.

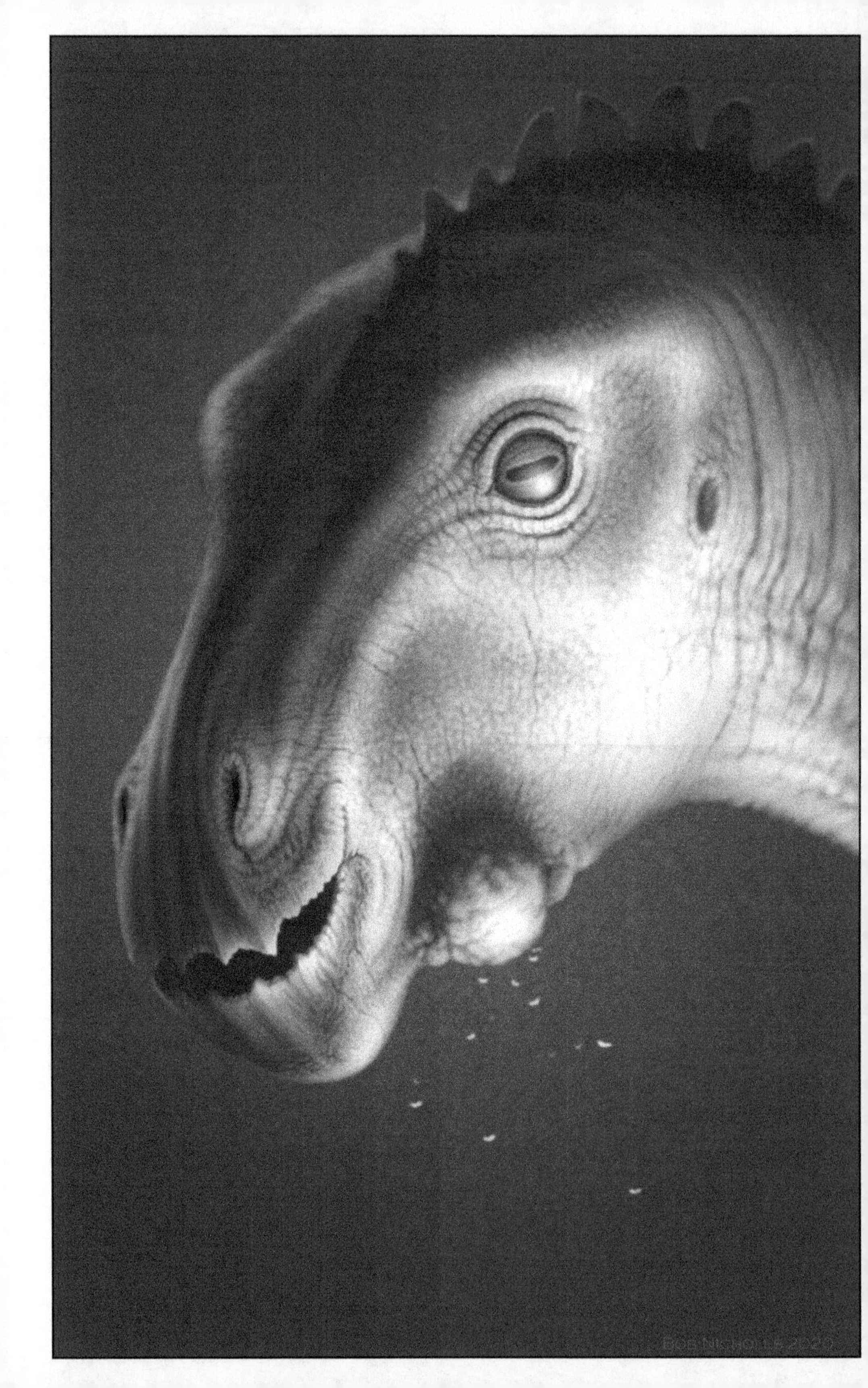

This individual is about half the size of the largest known *Telmatosaurus* specimens and was not fully mature when it died. Unfortunately, the cause of death could not be determined due to the incomplete nature of the fossil. However, even though ameloblastoma is benign, its uncontrolled growth can lead to severe, potentially debilitating complications that in all likelihood contributed to the demise of this young *Telmatosaurus*. Consequently, although the tumor had not developed into an enormous size, it would have been noticeable enough among healthy members of its herd that they could have inadvertently singled it out. In a modern setting, given the choice, predators often attack the young, elderly, and weak rather than prime adults. In one example, predatory seabirds called Skua are known to target and prey on Antarctic penguin chicks with malformations (those that are noticeably sick or different).

Tumors, cancers, and similar conditions found in humans and other animals have been affecting dinosaurs and various prehistoric species for millions of years. Understanding how living species respond to and cope with the side effects of such afflictions can help to reveal more about the lives and behaviors of these long-extinct animals.

FIGURE 5.14. (Opposite). *A portrait of malignance.*

A *Telmatosaurus transsylvanicus* with a noticeably enlarged left lower jaw resulting from the presence of an ameloblastoma tumor.

FOSSILIZED "FARTS"

Flatulence, passing wind, or farting—call it what you like. We all do it . . . a lot. Sometimes it can even clear a room. This gas is produced during digestion, generally in the stomach and/or gut, and is released from the anus; its smell and frequency varies depending on an individual's diet, health, and gut flora. Although we tend to think of farting as a human behavior, many animals do it, too, as my dog constantly proves. In like case, animals surely must have been farting for millions of years.

You are probably thinking, understandably, "Where are you going with this? How in the prehistoric world could we even have evidence of ancient farts?" Now that you are thinking about fossil farts, as bizarre as that might be, it is fair to assume that dinosaurs spring to mind, and now you're curious about whether a *T. rex* let rip. Well, considering that all of the 10,000+ birds (living theropods) do not fart, it can be assumed that rex probably didn't either. Though, not to disappoint, the plant-munching machines like giant sauropods and ornithischians (hadrosaurs, stegosaurs, ceratopsians, etc.) conceivably were very gassy, just as are large herbivores today. The only direct proof we have from fossils comes from something much smaller than dinosaurs but no less intriguing: insects in amber.

Yes, insects fart, although not all of them. Of those that have been studied by flatologists, the name for a scientist who studies flatulence (aka flatology), a beaded lacewing species called *Lomamyia latipennis* has evolved one of the deadliest of all farts. During its larval stage, it farts in the face of termites. This might not sound all that deadly, but a few minutes later, the termite becomes paralyzed from the toxic blast, and the larva begins to eat its fart-induced prey!

Insects are by far the most common animals to be found in amber. As the insects became trapped by the sticky resin that oozed from trees, their multitude of behaviors have been captured in real time, including various forms of mating, feeding, and parasitism, among many others. Gas bubbles are frequently found in amber as well and are sometimes associated with insects, usually positioned under their wings or legs. In most cases, the bubbles represent ancient air that was incorporated into the resin as it flowed down the tree, although some of the air was introduced by the insects when they either landed on or struggled to escape from the resin.

In rare instances, the bubbles emerged directly from the insect's rectum, as the result of a fart. Evidence supporting this interpretation includes the position of the bubbles at the insect's anus, the fine and intact preservation of the insect, and the lack of any general decomposition. Despite their rarity, numerous definite examples are known from both Baltic (Eocene age, 44–49 million years old) and Dominican (Miocene age, 16–20 million years old) amber and contain termites, cockroaches, ants, bees, beetles, and flies, all passing gas. It has also been claimed that some of the bubbles contain gases found in farts, such as methane and carbon dioxide.

It is amusing to say that we have fossil farts that are millions of years old, but there is more to it than that. Although these particular insects must have gotten stuck right before their bodies were readying the release of gas, they were not necessarily caught letting one go. Rather, after they became trapped in the resin and died, probably in a matter of seconds, tiny microbes remained alive long enough inside their gut to continue to digest

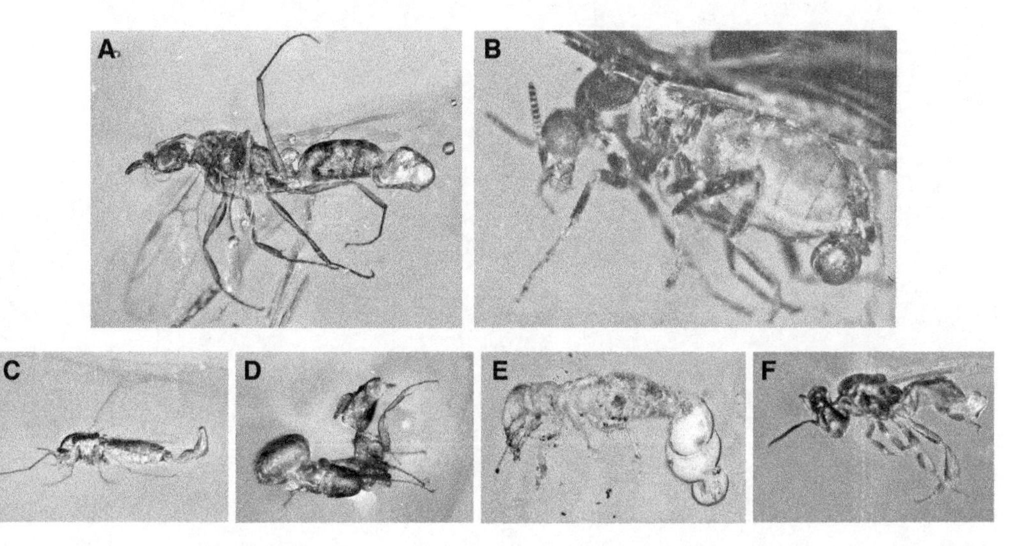

FIGURE 5.15. Insects with gas bubbles (flatus) escaping from their anus. (*A*) Flying ant. (*B*) Female blood-sucking blackfly in 40-million-year-old Baltic amber. (*C*) Midge. (*D*) Worker ant. (*E*) Worker termite with three separate passings of flatus. (*F*) Stingless bee. Figures 5.15 A, 5.15 C–F are contained in Dominican amber.

(Courtesy of George Poinar)

BOB NICHOLLS 2020

the animal's last meal, which led to the expulsion of gas (a fart), represented by the gas bubbles. Similar breaking down of food and releasing of gas after death is known to occur in several modern insects, including termites, which have enormous numbers of gut microbes, many of which are anaerobic. Even bloated termites are found in amber, caused by the build-up of gases that were unable to escape because the resin was too thick or the anus was blocked.

These fossilized farters capture one of the most unlikely behaviors to be preserved in the fossil record. They provide indirect evidence of insect–intestinal microbial associations, representing a mutualistic (beneficial) symbiotic relationship. The oldest direct evidence of insects and their microbes entombed in amber dates from at least 99 million years. These fossils confirm that ancient insects passed gas as a result of microorganism activity in their gut, as is the case among insects and other animals today, including humans. Their discovery takes the meaning of "silent but deadly" to a whole different, multi-million-year-old level.

FIGURE 5.16. (Opposite). *Stuck in time.*

Gooey resin drips down the trunk of a conifer tree and catches insects in its path, including an unfortunate flatulent termite.

COULD IT BE DINOSAUR PEE?

Hang on a minute, the last was fossil farts, and now dinosaur pee. Surely this is a joke?

Granted, the thought of dinosaurs urinating sounds more like a strange (although typical) conversation shared among paleontologists, but I assure you that this is based on science. After all, dinosaurs had to pee, too, so why should we not find evidence of this common behavior?

We know that fossil feces (coprolites) are a thing, but so is fossil urine. It might seem bizarre, but pee impressions can fossilize just like any other trace, given the right conditions. The name for these ancient traces is *urolite*, literally "urine stone" in Greek. The first report of a putative dinosaur urolite stems from research undertaken by paleontologists Katherine McCarville and Gale Bishop, whose findings made quite a splash at a professional paleontology conference in 2002.

What the pair had discovered, in their words, was a "bathtub-shaped depression" surrounded by hundreds of dinosaur tracks situated along the Purgatoire river, south of the town of La Junta, Colorado. A famous dinosaur track site, this Late Jurassic locality represents a 150-million-year-old lakeshore that is filled with tracks left by numerous sauropods and theropods. The unusual depression measured approximately 3 meters long, 1.5 meters wide, and 25–30 centimeters deep. In an attempt to replicate the trace, McCarville and Bishop performed experiments by streaming water across sand, which resulted in similar splats. At the track site, there are no signs of overhanging cliffs or other structures that water could have dropped from. Therefore, they hypothesized that the only explanation for this large, fluid-induced depression was the expulsion of urine from one of the dinosaurs, most probably a sauropod (something like *Diplodocus*).

Some might say that this idea of dinosaur pee is too far a stretch. And, in fact, work on this purported dinosaur urolite is still waiting to be formally published. That said, in 2004, two smaller, oval-shaped urolites were described in detail, collected from a quarry in Brazil's Paraná Basin (São Paulo State). Recorded from roughly 130-million-year-old Early Cretaceous rocks, the specimens were found in a fossilized sand dune deposit and were once again associated with dinosaur footprints, this time belonging to theropods and ornithopods.

Both traces possess a distinct crater-like pit where the liquid first hit the dry sand with force, which then trickled down a gentle slope, leaving behind ripple-like streamlines. To understand how the fossil structures were formed, the research team, led by Marcelo Fernandes, performed a simple experiment by pouring 2 liters of water from a height of 80 centimeters onto a sloped area of loose sand. This resulted in the formation of a structure similar to the fossils.

To further clarify their findings, the team compared the traces with those produced by an ostrich, the largest living dinosaur. At this point, it is important to note that birds do not pee like we do. Their liquid and solid waste is removed via a single opening called the *cloaca*, a structure that is

FIGURE 5.17. (*A*) One of the urolites attributed to a dinosaur from Brazil's Paraná Basin. The crater-like pit indicates where the liquid waste first hit with force and left behind ripple-like marks. (*B*) Modern ostrich with powerful pee hitting the ground with force.

(Courtesy of Marcelo Adorna Fernandes)

BOB NICHOLLS 2020

also present in crocodilians and has even been found in one extremely rare specimen of the dinosaur *Psittacosaurus* (first identified by a team including Bob Nicholls, the artist of this book). However, although many birds expel all of their waste at the same time, ostriches and other ratites flush out their fluid before ejecting their feces. The ostrich's powerful pee traces left behind in the dirt match the fossil structures, supporting the idea that these prehistoric traces were created by dinosaurs peeing in the sand.

Never say "never," but the chance of finding a dinosaur fossil in the act of peeing is pretty remote, to say the least. Consequently, besides birdwatching, urolites such as these are probably the closest things we will ever get to a peeing dinosaur.

As a final thought, I got to thinking about elephants peeing (and pooing, for that matter) and what this meant for the enormous sauropods. You can only imagine what it might have been like if you were a small animal standing in the most unfortunate place, right beneath one of these giants as it relieved itself. Perhaps this even led to death? What a way to go . . .

FIGURE 5.18. (Opposite). *The giant's relief.*

An explosion of pee is unleashed from a *Diplodocus*, causing mass panic among a small herd of tiny dinosaurs called *Fruitadens*. They scamper to safety in fear of their lives.

Acknowledgments

DEAN LOMAX

My mum, Anne Lomax, has been the biggest supporter of my career, sacrificing so much, giving constant love and support, and encouraging me every step along the way, from my dino-mad childhood to my career today. It is excruciatingly painful to say that during my time writing this book, my mum passed away. She was the kindest, most wonderful and generous person, truly everything and more that you could ever ask for in a parent. Mum didn't go to college, and our family had only enough money to just about pay the bills each month, yet Mum gave everything she could to help me and my brother and sister. She always wanted to spend time with her family, make everyone smile, and create happy memories. I am immensely proud to call her my mum.

Mum had been there at every step of my career, from waving me goodbye when I embarked on my career-defining trip to Wyoming to helping at book signings and seeing me get my PhD in 2019. She came to as many of my events as possible, listened to my public lectures, helped to promote my books, and she would race to tell all of her friends when I was on TV and radio. She even came fossil hunting with me. No amount of words can express how much I will miss her.

Everything I have ever achieved I put down to my mum. I wouldn't be the person I am today if it wasn't for her. This book is for you, Mum. Thank you for being *you*. I love you.

To the rest of my immediate family, I would like to say a huge thanks for your constant love and support and for putting up with me while I wrote this book: Joyce Lightfoot; Scott and Ken Lomax; Julie, Mark, Olivia, and Fletcher Boyles; Reece Davies; and Natalie Turner. A huge additional thanks to Natalie, who kindly edited and commented on the first draft of this book and has been incredibly supportive of everything that I do.

I express my sincere thanks to my good friend and fellow paleontologist Jason Sherburn(!), who kindly reviewed the book and provided invaluable comments that greatly improved it. He also happens to be the cohost of our paleontology podcast, *On the Fossil Record*. To my close friends and colleagues Nigel Larkin, Judy Massare, and Matt Holmes, it has been a highlight of my career getting to know you and working with you over the past decade. Thank you for everything that you have done for me.

It is impossible for me not to acknowledge my good friends and colleagues at the Wyoming Dinosaur Center, who gave an 18-year-old dino-mad kid from Doncaster, England, a chance at making his dreams become a reality. It was my first trip to Wyoming in 2008 that formed the backbone of my future career and which ultimately led to the study of the horseshoe crab death track that inspired the creation of this book.

Additional thanks to Victoria Arbour, Spencer Lucas, and an anonymous reviewer, who each provided very kind and complimentary reviews and offered some excellent suggestions to help improve parts of the book.

This book is the culmination of many people's work. It is based on countless years of dedicated research undertaken by fellow paleontologists. To each and every one of you, I say thank you—thank you for following your passion and helping to bring to life some of the most remarkable fossils ever found. In particular, I would like to acknowledge the work of Arthur Boucot, who dedicated much of his academic career to studying and compiling research on fossil behavior. In 1990, he introduced the term "frozen behaviour" for those fossil organisms preserved while actually in the midst of doing something.

I am very grateful to the following people, for you having kindly taken the time to chat with me about your research and discoveries, offer words of encouragement, and provide information that has helped to make this book a reality: Paul Barrett, Malcolm Bedell Jr., Mike Benton, Robert Boessenecker, Daniel Brinkman, Steve Brusatte, Markus Bühler, Steve Etches, Mike Everhart, Andy Farke, Marcelo Adorna Fernandes, Brian Fernando, Heinrich Frank, Mark Graham, Angie Guyon and the Wyoming Dinosaur Center, Lee and Ashley Hall, Ellie Harrison, Rolf Hauff, Dave Hone, Elaine Howard, ReBecca Hunt-Foster, Jim Kirkland, Adiël Klompmaker, Jessica Lippincott, Kristen MacKenzie, Susie Maidment, Andrea Marshall, Tony Martin, Levy and Kelly Morrow, Darren Naish, John Nudds, Conni O'Connor, Elsa Panciroli, Susan Passmore, David Penney, John Pickrell, Nick Pyenson, John Robinson, Andrew Rossi, Sven Sachs, Ross Secord, Tom Sermon, Levi Shinkle, Aaron Smith, Krister Smith, Hans-Dieter Sues, Jon Tennant, Mike Triebold, Jack Tseng, Bill Wahl, Betty and Warren Withers, Mark Witton, the Western Interior Paleontological Society (WIPS), and Zhonghe Zhou. I am also extremely grateful to those who have been kind enough to allow the use of photos in this book—thank you very much.

Finally, an enormous thanks to the amazingly talented Bob Nicholls for being an all-around brilliant and super-talented artist, scientist, and friend, but especially for bringing the fossils, their stories, and this book to life in such an extraordinary way.

BOB NICHOLLS

This book was Dean's idea, so my first acknowledgment is to him for inviting me to be a part of it. I pursued a career in paleo-reconstruction art because I fell in love with the illustrations of prehistoric animals in my childhood books and wanted to see my own paintings on bookshelves, too. After twenty years of painting very dead things for a living, the thrill of being published has never diminished, and this book is special because it is only the second time that I have had the opportunity to be the sole illustrator of a book from cover to cover. This project has been a career highlight, so, thanks, Dean. I sincerely hope there are many more to come!

My name is Bob Nicholls, and I am a workaholic. I work a lot, I paint and build models day and night, to the point that I consider the necessity of sleep an inconvenience. My wife, Victoria, endures this unreasonable fanaticism with a level of patience and understanding that amazes me. I am so grateful that she puts up with my paleoart obsession, but I am most grateful of all that she said “yes” on Durdle Door beach. Thanks, Victoria, you are the absolute best.

Thanks to my daughters, Darcey and Holly, for making every day an exhausting playtime. You make happy smiling my default setting. I will always be happy to pause work for “daddles” (dad cuddles) and I hope this book makes you proud enough to want to take it to school for show and tell.

I work in isolation and constantly rely on the expert publications of hundreds of scientists around the world to reconstruct my subjects. Their dedication and professionalism enable me to create plausible and up-to-date renderings of long-extinct creatures. There are too many to thank individually, but the papers and books of Darren Naish, Mark Witton, Scott Hartman, Gregory Paul, and Jakob Vinther are particularly influential for me. I urge all budding paleoartists to follow their work.

Finally, thanks to my parents for everything, all the time. You helped me do what I do, and what I do makes my life wonderful. I dedicate the pictures in this book to you.

• • •

We would both like to say “thank you” to our fabulous agent, Ariella Feiner, and to Molly Jamieson, who listened to the ideas of two enthusiastic paleontology geeks and guided us on a route that made this book a real thing that we can hold in our hands. To Miranda Martin and everybody at Columbia University Press, thank you for all the support and encouragement and for making this process so very enjoyable.

Further Reading

Attenborough, D. F. 1990. *The Trials of Life: A Natural History of Animal Behaviour*. London: Collins/BBC, 320.

Bottjer, D. J. 2016. *Paleoecology: Past, Present and Future*. Chichester, UK: Wiley, 232.

Boucot, A. J. 1990. *Evolutionary Paleobiology of Behavior and Coevolution*. New York: Elsevier, 725.

Boucot, A. J., and G. O. Poinar Jr. 2010. *Fossil Behavior Compendium*. Boca Raton, FL: CRC, 424.

Drickamer, L. C., S. H. Vessey, and E. M. Jakob. 2002. *Animal Behavior: Mechanisms, Ecology, Evolution*, 5th ed. Boston: McGraw-Hill, 422.

Hone, D. W. E., and C. G. Faulkes. 2013. "A Proposed Framework for Establishing and Evaluating Hypotheses About the Behaviour of Extinct Organisms." *Journal of Zoology* 292: 260–67.

Martin, A. J. 2014. *Dinosaurs Without Bones: Dinosaur Lives Revealed by Their Trace Fossils*. New York: Pegasus, 460.

Martin, A. J. 2017. *The Evolution Underground: Burrows, Bunkers, and the Marvelous Subterranean World Beneath Our Feet*. New York: Pegasus, 405.

INTRODUCTION—UNLOCKING THE PREHISTORIC WORLD

Benton, M. J. 2010. "Studying Function and Behavior in the Fossil Record." *PLoS Biology* 8, e1000321.

Boucot, A. J. 1990. *Evolutionary Paleobiology of Behavior and Coevolution*. New York: Elsevier, 725.

Boucot, A. J., and G. O. Poinar Jr. 2010. *Fossil Behavior Compendium*. Boca Raton, FL: CRC, 424.

Peckre, L. R., et al. 2018. "Potential Self-Medication Using Millipede Secretions in Red-Fronted Lemurs: Combining Anointment and Ingestion for a Joint Action Against Gastrointestinal Parasites?" *Primates* 59: 483–94.

SEX

Introduction

Bondar, C. 2015. *The Nature of Sex: The Ins and Outs of Mating in the Animal Kingdom*. London: W&N, 377.

Fisher, D. O., et al. 2013. "Sperm Competition Drives the Evolution of Suicidal Reproduction in Mammals." *PNAS* 110: 17910–914.

Knell, R. J., et al. 2013. "Sexual Selection in Prehistoric Animals: Detection and Implications." *Trends in Ecology & Evolution* 28: 38–47.

Mother Fish and Sex as We Know It

Long, J. A. 2012. *The Dawn of the Deed: The Prehistoric Origins of Sex*. Chicago: University of Chicago Press, 278.

Long, J. A. 2014. "Copulate to Populate: Ancient Scottish Fish Did It Sideways." *The Conversation*, October 19, 2014. https://theconversation.com/copulate-to-populate-ancient-scottish-fish-did-it-sideways-30910.

Long, J. A., et al. 2008. "Live Birth in the Devonian Period." *Nature* 453: 650–52.

Long, J. A., et al. 2015. "Copulation in Antiarch Placoderms and the Origin of Gnathostome Internal Fertilization." *Nature* 517: 196–99.

Newman, M. J., et al. 2020. "Earliest Vertebrate Embryos in the Fossil Record (Middle Devonian, Givetian)." *Palaeontology* 10.1111/pala.12511.

Shubin, N. 2009. *Your Inner Fish: The Amazing Discovery of Our 375-Million-Year-Old Ancestor*. London: Penguin, 256.

Dinosaur Sex Dance

Lockley, M. G., et al. 2016. "Theropod Courtship: Large Scale Physical Evidence of Display Arenas and Avian-Like Scrape Ceremony Behaviour by Cretaceous Dinosaurs." *Scientific Reports* 6, 18952.

Life in Death

Benton, M. J. 1991. "The Myth of the Mesozoic Cannibals." *New Scientist* 10: 40–44.

Böttcher, R. 1990. "Neue Erkenntnisse über die Fortpflanzungsbiologie der Ichthyosaurier (Reptilia)." *Stuttgarter Beiträge zur Naturkunde*, Serie B, 164: 1–51.

Chaning Pearce, J. 1846. "Notice of What Appears To Be the Embryo of an Ichthyosaurus in the Pelvic Cavity of *Ichthyosaurus (communis?)*." *Annals & Magazine of Natural History* 17: 44–46.

McGowan, C. 1979. "A Revision of the Lower Jurassic Ichthyosaurs of Germany, with the Description of Two New Species." *Palaeontographica Abt A* 166: 93–135

Motani, R., et al. 2014. "Terrestrial Origin of Viviparity in Mesozoic Marine Reptiles Indicated by Early Triassic Embryonic Fossils." *PLOS One* 9, e88640.

Jurassic Sex

Cryan, J. R., and G. J. Svenson. 2010. "Family-Level Relationships of the Spittlebugs and Froghoppers (Hemiptera: Cicadomorpha: Cercopoidea)." *Systematic Entomology* 35: 393–415.

Li, S., et al. 2013. "Forever Love: The Hitherto Earliest Record of Copulating Insects from the Middle Jurassic of China." *PLOS One* 8, e78188.

A Pregnant Plesiosaur

O'Keefe, F. R., and L. M. Chiappe. 2011. "Viviparity and K-Selected Life History in a Mesozoic Marine Plesiosaur (Reptilia, Sauropterygia)." *Science* 333: 870–73.

O'Keefe, F. R., et al. 2018. "Ontogeny of Polycotylid Long Bone Microanatomy and Histology." *Integrative Organismal Biology* 1, oby007.

Witton, M. P. 2019. "Plesiosaurs on the Rocks: The Terrestrial Capabilities of Four-Flippered Marine Reptiles." Markwittonblog.com (blog), January 25, 2019. http://markwitton-com.blogspot.com/2019/01/plesiosaurs-on-rocks-terrestrial.html.

When Whales Gave Birth on Land

Gingerich, P. D., et al. 2009. "New Protocetid Whale from the Middle Eocene of Pakistan: Birth on Land, Precocial Development, and Sexual Dimorphism." *PLOS One* 4, e4366.

Pyenson, N. D. 2017. "The Ecological Rise of Whales Chronicled by the Fossil Record." *Current Biology* 27: R558–R564.

Thewissen, J. G. M., et al. 2009. "From Land to Water: The Origin of Whales, Dolphins, and Porpoises." *Evolution: Education and Outreach* 2: 272–88.

Sexing a Cretaceous Bird

Chiappe, M. L. M., et al. 1999. "Anatomy and Systematics of the Confuciusornithidae (Theropoda: Aves) from the Late Mesozoic of Northeastern China." *Bulletin of the American Museum of Natural History* 242: 89.

Chinsamy, A., et al. 2013. "Gender Identification of the Mesozoic Bird *Confuciusornis sanctus*." *Nature Communications* 4: 1381.

Chinsamy, A., et al. 2020. "Osteohistology and Life History of the Basal Pygostylian, *Confuciusornis sanctus*." *Anatomical Record* 303: 949–62.
Hou, L-H., et al. 1995. "A Beaked Bird from the Jurassic of China." *Nature* 377: 616–18.
Li, Q., et al. 2018. "Elaborate Plumage Patterning in a Cretaceous Bird." *PeerJ* 6, e5831.
O'Connor, J. K., et al. 2014. "The Histology of Two Female Early Cretaceous Birds." *Vertebrata Palasiatica* 52: 112–28.
Varricchio, D. J., and F. D. Jackson. 2016. "Reproduction in Mesozoic Birds and Evolution of the Modern Avian Reproductive Mode." *The Auk* 133: 654–84.

Shell-Shocked Mating Turtles

Joyce, W. G., et al. 2012. "Caught in the Act: The First Record of Copulating Fossil Vertebrates." *Biology Letters* 8: 846–48.

Tiny Horses and Tiny Foals

Franzen, J. L. 2006. "A Pregnant Mare with Preserved Placenta from the Middle Eocene Maar of Eckfeld, Germany." *Palaeontographica Abteilung A* 278: 27–35.
Franzen, J. L. 2010. *The Rise of Horses: 55 Million Years of Evolution*. Baltimore: John Hopkins University Press, 211.
Franzen, J. L. 2017. "Report on the Discovery of Fossil Mares with Preserved Uteroplacenta from the Eocene of Germany." *Fossil Imprint* 73: 67–75.
Franzen, J. L., et al. 2015. "Description of a Well Preserved Fetus of the European Eocene Equoid *Eurohippus messelensis*." *PLOS One* 10, e0137985.

PARENTAL CARE AND COMMUNITIES

Introduction

Grone, B. P., et al. 2012. "Food Deprivation Explains Effects of Mouthbrooding on Ovaries and Steroid Hormones, but Not Brain Neuropeptide and Receptor mRNAs, in an African Cichlid Fish." *Hormones and Behavior* 62: 18–26.
Royle, N. J., et al. 2013. "Burying Beetles." *Current Biology* 23: R907–R909.
Schäfer, M., et al. 2019. "Goliath Frogs Build Nests for Spawning—The Reason for Their Gigantism?" *Journal of Natural History* 53: 1263–76.
Scott, M. P. 1990. "Brood Guarding and the Evolution of Male Parental Care in Burying Beetles." *Behavior Ecology and Sociobiology* 26: 31–39.
Scott, M. P. 1998. "The Ecology and Behaviour of Burying Beetles." *Annual Review of Entomology* 43: 595–618.
Shukla, S. P., et al. 2018. "Microbiome-Assisted Carrion Preservation Aids Larval Development in a Burying Beetle." *PNAS* 115: 11274–279.

Brooding Dinosaurs

Bi, S., et al., 2020. "An Oviraptorid Preserved Atop an Embryo-Bearing Egg Clutch Sheds Light on the Reproductive Biology of Non-Avialan Theropod Dinosaurs." *Science Bulletin*, doi.org/10.1016/j.scib.2020.12.018.

Clark, J. M., et al. 1999. "An Oviraptorid Skeleton from the Late Cretaceous of Ukhaa Tolgod, Mongolia, Preserved in an Avianlike Brooding Position Over an Oviraptorid Nest." *American Museum Novitates* 3265: 1–36.

Dong, Z-M., and P. J. Currie. 1996. "On the Discovery of an Oviraptorid Skeleton on a Nest of Eggs at Bayan Mandahu, Inner Mongolia, People's Republic of China." *Canadian Journal of Earth Science* 33: 631–36.

Erickson, G. M., et al. 2007. "Growth Patterns in Brooding Dinosaurs Reveals the Timing of Sexual Maturity in Non-Avian Dinosaurs and Genesis of the Avian Condition." *Biology Letters* 3: 558–61.

Fanti, F., et al. 2012. "New Specimens of *Nemegtomaia* from the Baruungoyot and Nemegt Formations (Late Cretaceous) of Mongolia." *PLOS One* 7, e31330.

Norell, M. A., et al. 1994. "A Theropod Dinosaur Embryo and the Affinities of the Flaming Cliffs Dinosaur Eggs." *Science* 266: 779–82.

Norell, M. A., et al. 1995. "A Nesting Dinosaur." *Nature* 378: 774–76.

Norell, M. A., et al. 2018. "A Second Specimen of *Citipati osmolskae* Associated with a Nest of Eggs from Ukhaa Tolgod, Omnogov Aimag, Mongolia." *American Museum Novitates* 3899: 1–44.

Osborn, H. F. 1924. "Three New Theropod, *Protoceratops* Zone, Central Mongolia." *American Museum Novitates* 144: 1–12.

Tanaka, K., et al. 2018. "Incubation Behaviours of Oviraptorosaur Dinosaurs in Relation to Body Size." *Biology Letters* 14, 20180135.

Varricchio, D. J., et al. 2008. "Avian Paternal Care Had Dinosaur Origin." *Science* 322: 1826–28.

Yang, T.-R., et al. 2019. "Reconstruction of Oviraptorid Clutches Illuminates Their Unique Nesting Biology." *Acta Palaeontologica Polonica* 64: 581–96.

Oldest Parental Care

Caron, J.-B., and J. Vannier. 2016. "*Waptia* and the Diversification of Brood Care in Early Arthropods." *Current Biology* 26: 69–74.

Duan, Y., et al. 2014. "Reproductive Strategy of the Bradoriid Arthropod *Kunmingella douvillei* from the Lower Cambrian Chengjiang Lagerstätte, South China." *Gondwana Research* 25: 983–90.

Fu, D., et al. 2018. "Anamorphic Development and Extended Parental Care in a 520-Million-Year-Old Stem-Group Euarthropod from China." *BMC Evolutionary Biology* 18: 139–48.

Vannier, J., et al. 2018. "*Waptia fieldensis* Walcott, a Mandibulate Arthropod from the Middle Cambrian Burgess Shale." *Royal Society Open Science* 5, 172206.

Pterosaur Nesting Grounds

Chiappe, L. M., et al. 2004. "Argentinian Unhatched Pterosaur Fossil." *Nature* 432: 571–72.

Ji, Q., et al. 2004. "Pterosaur Egg with a Leathery Shell." *Nature* 432: 572.

Lű, J., et al. 2011. "An Egg-Adult Association, Gender, and Reproduction in Pterosaurs." *Science* 331: 321–24.

Unwin, D. M., and C. D. Deeming. 2019. "Prenatal Development in Pterosaurs and Its Implications for Their Postnatal Locomotory Ability." *Proceedings of the Royal Society B* 286: 20190409.

Wang, X., and Z. Zhou. 2004. "Pterosaur Embryo from the Early Cretaceous." *Nature* 429: 621.

Wang, X., et al. 2014. "Sexually Dimorphic Tridimensionally Preserved Pterosaurs and Their Eggs from China." *Current Biology* 24: 1323–30.

Wang, X., et al. 2015. "Eggshell and Histology Provide Insight on the Life History of a Pterosaur with Two Functional Ovaries." *Anais da Academia Brasileira de Ciências* 87: 1599–1609.

Wang, X., et al. 2017. "Egg Accumulation with 3D Embryos Provides Insight Into the Life History of a Pterosaur." *Science* 358: 1197–1201.

Megalodon Nursery

Boessenecker, R. W., et al. 2019. "The Early Pliocene Extinction of the Mega-Toothed Shark *Otodus megalodon*: A View from the Eastern North Pacific." *PeerJ* 7, e6088.

Herraiz, J. L., et al. 2020. "Use of Nursery Areas by the Extinct Megatooth Shark *Otodus megalodon* (Chondrichthyes: Lamniformes)." *Biology Letters* 16: 20200746.

Heupel, M. R., et al. 2007. "Shark Nursery Areas: Concepts, Definition, Characterization and Assumptions." *Marine Ecology Progress Series* 337: 287–97.

Pimiento, C., et al. 2010. "Ancient Nursery Area for the Extinct Giant Shark Megalodon from the Miocene of Panama." *PLOS One* 5, e10552.

Shimada, K. 2019. "The Size of the Megatooth Shark, *Otodus megalodon* (Lamniformes: Otodontidae), Revisited." *Historical Biology*. https://doi.org/10.1080/08912963.2019.1666840

The Babysitter

Coombs, W. P. 1982. "Juvenile Specimens of the Ornithischian Dinosaur *Psittacosaurus*. *Palaeontology* 25: 89–107.

Erickson, G. M., et al. 2009. "A Life Table for *Psittacosaurus lujiatunensis*: Initial Insights Into Ornithischian Dinosaur Population Biology." *Anatomical Record* 292: 1514–21.

Hedrick, B. P., et al. 2014. "The Osteology and Taphonomy of a *Psittacosaurus* Bonebed Assemblage of the Yixian Formation (Lower Cretaceous), Liaoning, China." *Cretaceous Research* 51: 321–40.

Isles, T. E. 2009. "The Socio-Sexual Behaviour of Extant Archosaurs: Implications for Understanding Dinosaur Behaviour." *Historical Biology* 21: 139–214.

Meng, Q., et al. 2004. "Parental Care in an Ornithischian Dinosaur." *Nature* 431: 145–46.
Vinther, J., et al. 2016. "3D Camouflage in an Ornithischian Dinosaur." *Current Biology* 26: 2456–62.
Zhao, Q., et al. 2007. "Social Behaviour and Mass Mortality in the Basal Ceratopsian Dinosaur *Psittacosaurus* (Early Cretaceous, People's Republic of China)." *Palaeontology* 50: 1023–29.
Zhao, Q., et al. 2014. "Juvenile-Only Clusters and Behaviour of the Early Cretaceous Dinosaur *Psittacosaurus*." *Acta Palaeontologica Polonica* 59: 827–33.

Dinosaur Death Trap

Franz, A. 2017. "Locked in Rock: Liberating America's Giant Raptors." Earth Archives. http://www.eartharchives.org/articles/locked-in-rock-liberating-america-s-giant-raptors/.
Kirkland, J. I., et al. 1993. "A Large Dromaeosaur (Theropoda) from the Lower Cretaceous of Eastern Utah." *Hunteria* 2: 1–16.
Kirkland, J. I., et al. 2016. "Depositional Constraints of the Lower Cretaceous Stikes Quarry Dinosaur Site: Upper Yellow Cat Member, Cedar Mountain Formation, Utah." *Palaios* 31: 421–39.
Li, R., et al. 2008. "Behavioral and Faunal Implications of Early Cretaceous Deinonychosaur Trackways from China." *Naturwissenschaften* 95: 185–91.
Madsen, S. 2016. "The Utahraptor Project." https://www.gofundme.com/f/utahraptor.
Maxwell, D. W., and J. H. Ostrom. 1995. "Taphonomy and Paleobiological Implications of *Tenontosaurus-Deinonychus* Associations." *Journal of Vertebrate Paleontology* 15: 707–12.
Moscato, D. 2016. "What Killed the Dinosaurs in this Fossilised Mass Grave?" Earth Touch News Network, October 10, 2016. https://www.earthtouchnews.com/discoveries/fossils/what-killed-the-dinosaurs-in-this-fossilised-mass-grave/.
Roach, B. T., and D. L. Brinkman. 2007. "A Reevaluation of Cooperative Pack Hunting and Gregariousness in *Deinonychus antirrhopus* and Other Nonavian Theropod Dinosaurs." *Bulletin of the Peabody Museum of Natural History* 48: 103–38.

A Prehistoric Pompeii

"Ashfall Fossil Beds State Historical Park Designated a National Natural Landmark" (brochure). Nebraska Game and Parks Commission and the University of Nebraska State Museum. https://ashfall.unl.edu/file_download/inline/f8fed67f-c2ed-4a3d-adc3-47cb92b7d8cc.
Mead, A. J. 2000. "Sexual Dimorphism and Paleoecology in *Teleoceras*, a North American Miocene Rhinoceros." *Paleobiology* 26: 689–706.
Mihlbachler, M. C. 2003. "Demography of Late Miocene Rhinoceroses (*Teleoceras proterum* and *Aphelops malacorhinus*) from Florida: Linking Mortality and Sociality in Fossil Assemblages." *Paleobiology* 29: 412–28.
Tucker, S. T., et al. 2014. "The Geology and Paleontology of Ashfall Fossil Beds, a Late Miocene (Clarendonian) Mass-Death Assemblage, Antelope County and Adjacent Knox County, Nebraska, USA." *Geological Society of America Field Guide* 36: 1–22.

Voorhies, M. R. 1985. "A Miocene Rhinoceros Herd Buried in Volcanic Ash." *National Geographic Society Research Reports* 19: 671–88.

Voorhies, M. R., and S. G. Stover. 1978. "An Articulated Fossil Skeleton of a Pregnant Rhinoceros, *Teleoceras major* Hatcher." *Proceedings, Nebraska Academy of Sciences* 88: 47–48.

Voorhies, M. R., and J. R. Thomasson. 1979. "Fossil Grass Anthoecia Within Miocene Rhinoceros Skeletons: Diet in an Extinct Species." *Science* 206: 331–33.

Fish Trapped Inside Giant Clams

Kauffman, E. G., et al. 2007. "Paleoecology of Giant Inoceramidae (*Platyceramus*) on a Santonian (Cretaceous) Seafloor in Colorado." *Journal of Paleontology* 81: 64–81.

Neo, M. L., et al. 2015. "The Ecological Significance of Giant Clams in Coral Reef Ecosystems." *Biological Conservation* 181: 111–23.

Soo, P., and P. A. Todd. 2014. "The Behaviour of Giant Clams (Bivalvia: Cardiidae: Tridacninae)." *Marine Biology* 161: 2699–2717.

Stewart, J. D. 1990. "Niobrara Formation Symbiotic Fish in Inoceramid Bivalves." In *Society of Vertebrate Paleontology Niobrara Chalk Excursion Guidebook*, ed. S. Christopher Bennett, 31–41. Lawrence: Museum of Natural History and the Kansas Geological Survey.

Stewart, J. D. 1990. "Preliminary Account of Halecostome-Inoceramid Commensalism in the Upper Cretaceous of Kansas." In *Evolutionary Paleobiology of Behavior and Coevolution*, ed. A. J. Boucot, 51–57. Amsterdam: Elsevier.

Wiley, E. O., and J. D. Stewart. 1981. "*Urenchelys abditus*, New Species, the First Undoubted Eel (Teleostei: Anguilliformes) from the Cretaceous of North America." *Journal of Vertebrate Paleontology* 1: 43–47.

Snowmastodon

Black, R. 2014. "Laelaps, Snowsalamander." *National Geographic*, December 15, 2014. https://www.nationalgeographic.com/science/phenomena/2014/12/15/snowsalamander/

Johnson, K., and I. Miller. 2012. *Digging Snowmastodon. Discovering an Ice Age World in the Colorado Rockies*. Denver: Denver Museum of Nature & Science and People's Press, 141.

Sertich, J. J. W., et al. 2014. "High-Elevation Late Pleistocene (MIS 6–5) Vertebrate Faunas from the Ziegler Reservoir Fossil Site, Snowmass Village, Colorado." *Quaternary Research* 82: 504–17.

Giant Floating Ecosystem

Hauff, B., and R. B. Hauff. 1981. *Das Holzmadenbuch*. Fellbach, Germany: REPRO-DRUCK, 136.

Hess, H. 1999. *Lower Jurassic Posidonia Shale of Southern Germany*. In *Fossil Crinoids*, ed. H. Hess et al., 183–96. Cambridge: Cambridge University Press.

Hess, H. 2010. "Paleoecology of Pelagic Crinoids." *Treatise Online* 16: 1–33.

Hunter, A. W., et al. 2020. "Reconstructing the Ecology of a Jurassic Pseudoplanktonic Raft Colony." *Royal Society Open Science* 7: 200142.

Seilacher, A. 1968. "Form and Function of the Stem in a Pseudoplanktonic Crinoid (*Seirocrinus*)." *Palaeontology* 11: 275–82.

Seilacher, A., and R. B. Hauff. 2004. "Constructional Morphology of Pelagic Crinoids." *Palaios* 19: 3–16.

Thiel, M., and C. Fraser. 2016. "The Role of Floating Plants in Dispersal of Biota Across Habitats and Ecosystems." In *Marine Macrophytes as Foundation Species*, ed. E. Olafsson, 76–99. Boca Raton, FL: CRC.

MOVING AND MAKING HOMES

Introduction

Huffard, C. L., et al. 2005. "Underwater Bipedal Locomotion by Octopuses in Disguise." *Science* 307: 1927.

Smith, T. S., et al. 2013. "An Improved Method of Documenting Activity Patterns of Post-Emergence Polar Bears (*Ursus maritimus*) in Northern Alaska." *Arctic* 66: 139–46.

Steyn, P. 2017. "How Does the Great Wildebeest Migration Work?" *National Geographic*, February 8, 2017. https://blog.nationalgeographic.org/2017/02/08/how-does-the-great-wildebeest-migration-work/.

Wilson, N. 2015. "Why Termites Build Such Enormous Skyscrapers." BBC Earth (December 15). http://www.bbc.com/earth/story/20151210-why-termites-build-such-enormous-skyscrapers.

Mammal Movers

Martill, D. M. 1988. "Flash Floods and Panic in the Fossil Record: A Tale of 25 Titanotheres." *Geology Today* 4: 27–30.

McCarroll, S. M., et al. 1996. *The Mammalian Faunas of the Washakie Formation, Eocene Age, of Southern Wyoming. Part III. The Perissodactyls.* Fieldiana 33. Chicago: Field Museum of Natural History, 38.

Mihlbachler, M. C. 2008. *Species Taxonomy, Phylogeny, and Biogeography of the Brontotheriidae (Mammalia: Perissodactyla).* In *Bulletin of the American Museum of Natural History* 311. New York: American Museum of Natural History.

Turnbull, W. D., and D. M. Martill. 1988. "Taphonomy and Preservation of a Monospecific Titanothere Assemblage from the Washakie Formation (Late Eocene), Southern Wyoming. An Ecological Accident in the Fossil Record." *Palaeogeography, Palaeoclimatology, Palaeoecology* 63: 91–108.

Follow the Leader

Błażejowski, B., et al. 2016. "Ancient Animal Migration: A Case Study of Eyeless, Dimorphic Devonian Trilobites from Poland." *Palaeontology* 59: 743–51.

Lawrance, P., and S. Stammers. 2014. *Trilobites of the World. An Atlas of 1000 Photographs.* Manchester, UK: Siri Scientific, 416.

Radwańksi, A., et al. 2009. "Queues of Blind Phacopid Trilobites *Trimerocephalus*: A Case of Frozen Behaviour of Early Famennian Age from the Holy Cross Mountains, Central Poland." *Acta Geologica Polonica* 59: 459–81.

Vannier, J., et al. 2019. "Collective Behaviour in 480-Million-Year-Old Trilobite Arthropods from Morocco." *Nature Scientific Reports* 9: 14941.

Xian-guang, H., et al. 2008. "Collective Behavior in an Early Cambrian Arthropod." *Science* 322: 224.

Xian-guang, H., et al. 2009. "A New Arthropod in Chain-Like Associations from the Chengjiang Lagerstätte (Lower Cambrian), Yunnan, China." *Palaeontology* 52: 951–61.

Sitting on the Jurassic Bay

Bird, R. T. 1985. *Bones for Barnum Brown: Adventures of a Dinosaur Hunter.* Fort Worth: Texas Christian University Press, 225.

Hitchcock, E. 1858. *Ichnology of New England: A Report on the Sandstone of the Connecticut Valley, Especially Its Fossil Footmarks.* Boston: William White, 232.

Lockley, M., et al. 2003. "Crouching Theropods in Taxonomic Jungles: Ichnological and Ichnotaxonomic Investigations of Footprints with Metatarsal and Ischial Impressions." *Ichnos* 10: 169–77.

Martin, A. J. 2014. *Dinosaurs Without Bones. Dinosaur Lives Revealed by Their Trace Fossils.* New York: Pegasus,. 460.

Milner, A. R. C., et al. 2009. "Bird-Like Anatomy, Posture, and Behavior Revealed by an Early Jurassic Theropod Dinosaur Resting Trace." *PLOS One* 4, e4591.

Pemberton, S. G., et al. 2007. "Edward Hitchcock and Roland Bird: Two Early Titans of Vertebrate Ichnology in North America." In *Trace Fossils: Concepts, Problems, Prospects,* ed. W. Miller III, 32–51. Burlington, MA: Elsevier Science.

Witton, M. P. 2016. "The Dinosaur Resting Pose Debate: Some Thoughts for Artists." Markwitton.com (blog), June 10, 2016. http://markwitton-com.blogspot.com/2016/06/the-dinosaur-resting-pose-debate-some.html.

The Death March

Lomax, D. R., and C. A. Racay. 2012. "A Long Mortichnial Trackway of *Mesolimulus walchi* from the Upper Jurassic Solnhofen Lithographic Limestone Near Wintershof, Germany." *Ichnos* 19: 189–97.

Mass Moth Migration

Ataabadi, M. M., et al. 2017. "A Locust Witness of a Trans-Oceanic Oligocene Migration Between Arabia and Iran (Orthoptera: Acrididae)." *Historical Biology* 31: 574–80.

Penney, D., and J. E. Jepson. 2014. *Fossil Insects. An Introduction to Palaeoentomology*. Manchester, UK: Siri Scientific, 222.

Rust, J. 2000. "Fossil Record of Mass Moth Migration." *Nature* 405: 530–31.

Sohn, J.-C., et al. 2015. "The Fossil Record and Taphonomy of Butterflies and Moths (Insecta, Lepidoptera): Implications for Evolutionary Diversity and Divergence-Time Estimates." *BMC Evolutionary Biology* 15: 12.

Dinosaur Death Pits

Eberth, D. A., et al. 2010. "Dinosaur Death Pits from the Jurassic of China." *Palaios* 25: 112–25.

Time to Grow When You Molt and Go

Daley, A. C., and H. B. Drage. 2016. "The Fossil Record of Ecdysis, and Trends in the Moulting Behaviour of Trilobites." *Arthropod Structure & Development* 45: 71–96.

Garcia-Bellido, D.C., and D. H. Collins. 2004. "Moulting Arthropod Caught in the Act." *Nature* 429: 40.

Giribet, G., and G. D. Edgecomb. 2019. "The Phylogeny and Evolutionary History of Arthropods." *Current Biology* 29: R592–R602.

Vallon, L. H., et al. 2015. "Ecdysichnia—A New Ethological Category for Trace Fossils Produced by Moulting." *Annales Societatis Geologorum Poloniae* 85: 433–44.

Yang, J., et al. 2019. "Ecdysis in a Stem-Group Euarthropod from the Early Cambrian of China." *Nature Scientific Reports* 9: 5709.

Prehistoric Partners

Fernandez, V., et al. 2013. "Synchrotron Reveals Early Triassic Odd Couple: Injured Amphibian and Aestivating Therapsid Share Burrow." *PLOS One* 8, e64978.

Jasinoski, S. C., and F. Abdala. 2017. "Aggregations and Parental Care in the Early Triassic Basal Cynodonts *Galesaurus planiceps* and *Thrinaxodon liorhinus*." *PeerJ* 5, e2875.

Smith, R. M. H. 1987. "Helical Burrow Casts of Therapsid Origin from the Beaufort Group (Permian) of South Africa." *Palaeogeography, Palaeoclimatology, Palaeoecology* 60: 155–70.

Devil's Corkscrews

Barbour, E. H. 1892. "Notice of New Gigantic Fossils." *Science* 19: 99–100.

Doody, J. S., et al. 2015. "Deep Nesting in a Lizard, *Déjà Vu* Devil's Corkscrews: First Helical Reptile Burrow and Deepest Vertebrate Nest." *Biological Journal of the Linnean Society* 116: 13–26.

Martin, L. D., and D. K. Bennett. 1977. "The Burrows of the Miocene Beaver *Palaeocastor*, Western Nebraska." *Palaeogeography, Palaeoclimatology, Palacoecology* 22: 173–93.

Meyer, R. C. 1999. "Helical Burrows as a Palaeoclimate Response: *Daimonelix* by *Palaeocastor*." *Palaeogeography, Palaeoclimatology, Palaeoecology* 147: 291–98.
Peterson, O. A. 1904. "Recent Observations Upon *Daemonelix*." *Science* 20: 344–45.
Peterson, O. A. 1905. "Description of New Rodents and Discussion of the Origin of *Daemonelix*." *Carnegie Museum Memoirs* 2: 139–202.
Sues, H.-D. 2019. "How Scientists Resolved the Mystery of the Devil's Corkscrews." *Smithsonian Magazine*, November 25, 2019. https://www.smithsonianmag.com/smithsonian-institution/how-scientists-resolved-mystery-devils-corkscrews-180973487/.

The Dinosaur That Lived in a Burrow

Fearon, J. L., and D. J. Varricchio. 2016. "Reconstruction of the Forelimb Musculature of the Cretaceous Ornithopod Dinosaur *Oryctodromeus cubicularis*: Implications for Digging." *Journal of Vertebrate Paleontology* 36, e1078341.
Krumenacker, L. J., et al. 2019. "Taphonomy of and New Burrows from *Oryctodromeus cubicularis*, a Burrowing Neornithischian Dinosaur, from the Mid-Cretaceous (Albian-Cenomanian) of Idaho and Montana, U.S.A." *Palaeogeography, Palaeoclimatology, Palaeoecology* 530: 300–11.
Martin, A. J. 2014. *Dinosaurs Without Bones. Dinosaur Lives Revealed by Their Trace Fossils.* New York: Pegasus, 460.
Varricchio, D. J., et al. 2007. "First Trace and Body Fossil Evidence of a Burrowing, Denning Dinosaur." *Proceedings of the Royal Society B* 274: 1361–68.

Giant Underground Sloths

Barlow, C. C. 2000. *The Ghosts of Evolution.* New York: Basic, 291.
Bell, C. M. 2002. "Did Elephants Hang from Trees?" *Geology Today* 18: 63–66.
Bustos, D., et al. 2018. "Footprints Preserve Terminal Pleistocene Hunt? Human-Sloth Interactions in North America." *Science Advances* 4, eaar7621.
Cliffe, B. 2016. "Sloths Aren't Lazy—Their Slowness Is a Survival Skill." *The Conversation*, April 19, 2016. https://theconversation.com/sloths-arent-lazy-their-slowness-is-a-survival-skill-63568.
Frank, H. T., et al. 2012. "Cenozoic Vertebrate Tunnels in Southern Brazil." In *Ichnology of Latin America: Selected Papers*, ed. R. G. Netto et al., 141–57. Monografias da Sociedade Brasileira de Paleontologia 2. Rio de Janiero: Sociedade Brasileira de Paleontologia.
Frank, H. T., et al. 2013. "Description and Interpretation of Cenozoic Vertebrate Ichnofossils in Rio Grande Do Sul State, Brazil." *Revista Brasileira de Paleontologia* 16: 83–96.
Frank, H. T., et al. 2015. "Underground Chamber Systems Excavated by Cenozoic Ground Sloths in the State of Rio Grande Do Sul, Brazil." *Revista Brasileira de Paleontologia* 18: 273–84.
Janzen, D. H., and P. S. Martin. 1982. "Neotropical Anachronisms: The Fruits the Gomphotheres Ate." *Science* 215: 19–27.

Lopes, R. P., et al. 2017. "*Megaichnus* igen. nov.: Giant Paleoburrows Attributed to Extinct Cenozoic Mammals from South America." *Ichnos* 24: 133–45.
Naish, D. 2005. "Fossils Explained 51: Sloths." *Geology Today* 21: 232–38.
Quinn, C. E. 1976. "Thomas Jefferson and the Fossil Record." *BIOS* 47: 159–67.
Vizcaino, S. F., et al. 2001. "Pleistocene Burrows in the Mar del Plata Area (Argentina) and Their Probable Builders." *Acta Palaeontologica Polonica* 46: 289–301.

FIGHTING, BITING, AND FEEDING

Introduction

Dorward, L. J. 2014. "New Record of Cannibalism in the Common Hippo, *Hippopotamus amphibius* (Linnaeus, 1758)." *African Journal of Ecology* 53: 385–87.
Jorgensen, S. J., et al. 2019. "Killer Whales Redistribute White Shark Foraging Pressure on Seals." *Nature Scientific Reports* 9: 6153.
Li, D., et al. 2012. "Remote Copulation: Male Adaptation to Female Cannibalism." *Biology Letters* 8: 512–15.
Pyle, P., et al. 1999. "Predation on a White Shark (*Carcharodon carcharias*) by a Killer Whale (*Orcinus orca*) and a Possible Case of Competitive Displacement." *Marine Mammal Science* 15: 563–68.

Clash of the Mammoths

Boucot, A. J. 1990. *Evolutionary Paleobiology of Behavior and Coevolution*. New York: Elsevier, 725.
Chelliah, K., and R. Sukumar. 2013. "The Role of Tusks, Musth and Body Size in Male-Male Competition Among Asian Elephants, *Elephas maximus*." *Animal Behaviour* 86: 1207–14.
Colyer, F., and A. E. W. Miles. 1957. "Injury to and Rate of Growth of an Elephant Tusk." *Journal of Mammalogy* 38: 243–47.
Fisher, D. C. 2004. "Season of Musth and Musth-Related Mortality in Pleistocene Mammoths." *Journal of Vertebrate Paleontology* 24: 58A.
Fisher, D. C. 2009. "Paleobiology and Extinction of Proboscideans in the Great Lakes Region of North America." In *American Megafaunal Extinctions at the End of the Pleistocene*, ed. G. Haynes, 55–75. Dordrecht: Springer.
Holen, S. R. 2006. "Taphonomy of Two Last Glacial Maximum Mammoth Sites in the Central Great Plains of North America: A Preliminary Report on La Sena and Lovewell." *Quaternary International* 142–143: 30–43.
Mol, D., et al. 2006. "The Yukagir Mammoth: Brief History, 14c Dates, Individual Age, Gender, Size, Physical and Environmental Conditions and Storage." *Scientific Annals* 98: 299–314.
PBS. 2008. *Mammoth Mystery* (documentary). *Nova* (July 30). https://www.pbs.org/wgbh/nova/nature/mammoth-mystery.html.

Poole, J. H. 1987. "Rutting Behavior in African Elephants: The Phenomenon of Musth." *Behaviour* 102: 283–316.

Poole, J. H., and C. J. Moss. 1981. "Musth in the African Elephant, *Loxodonta africana*." *Nature* 292: 830–31.

Rempp, K. 2012. "Clash of the Titans: 50 Years Later." *Rapid City Journal*, October 2, 2012. https://rapidcityjournal.com/community/chadron/clash-of-the-titans-50-years-later/article_addf527c-0cb3-11e2-8c4f-0019bb2963f4.html.

Voorhies, M. R. 1994. "Hooves and Horns." *Nebraska History* 75: 74–81.

The Fighting Dinosaurs

Barsbold, R. 1974. "Duelling Dinosaurs." *Priroda* 2: 81–83.

Barsbold, R. 2016. "The Fighting Dinosaurs: The Position of Their Bodies Before and After Death." *Palaeontological Journal* 50: 1412–17.

Barsbold, R. 2018. "On Morphological Diversity in Directed Development of Late Carnivorous Dinosaurs (Theropoda Marsh 1881)." *Paleontological Journal* 52: 1764–70.

Carpenter, K. 1998. "Evidence of Predatory Behaviour by Carnivorous Dinosaurs." *Gaia* 15: 135–44.

Hone, D., et al. 2010. "New Evidence for a Trophic Relationship Between the Dinosaurs *Velociraptor* and *Protoceratops*." *Palaeogeography, Palaeoclimatology, Palaeoecology* 291: 488–92.

Kielan-Jaworowska, Z., and R. Barsbold. 1972. "Narrative of the Polish-Mongolian Palaeontological Expeditions 1967–1971." *Palaeontologica Polonica* 27: 5–13.

Jurassic Drama

Frey, E., and H. Tischlinger. 2012. "The Late Jurassic Pterosaur *Rhamphorhynchus*, a Frequent Victim of the Ganoid Fish *Aspidorhynchus*?" *PLOS One* 7, e31945.

Weber, F. 2013. "Paléoécologie des ptérosaures 3: Les reptiles volants de Solnhofen, Allemagne." *Fossiles* 14: 50–59.

Witton, M. P. 2017. "Pterosaurs in Mesozoic Food Webs: A Review of Fossil Evidence." In *New Perspectives on Pterosaur Palaeobiology*, ed. D. W. E. Hone et al., 455. London: Geological Society.

Terror Worm of the Primeval Sea

Bruton, D. L. 2001. "A Death Assemblage of Priapulid Worms from the Middle Cambrian Burgess Shale." *Lethaia* 34: 163–67.

Conway-Morris, S. 1977. "Fossil Priapulid Worms." *Special Papers in Palaeontology* 20: 1–95.

Smith, M. R., et al. 2015. "The Macro- and Microfossil Record of the Cambrian Priapulid *Ottoia*." *Palaeontology* 58: 705–21.

Vannier, J. 2012. "Gut Contents as Direct Indicators for Trophic Relationships in the Cambrian Marine Ecosystem." *PLOS One* 7, e52200.

Wallace, R. L. 2002. "Priapulida." *Encyclopedia of Life Sciences*. New York: Wiley, 1–4.

Greedy Fish

Bardack, D. 1965. "Anatomy and Evolution of Chirocentrid Fishes." *Vertebrata* 10: 1–88.

Everhart, M. J. 2005. *Oceans of Kansas: A Natural History of the Western Interior Sea*. Bloomington: Indiana University Press, 322.

Everhart, M. J. 2010. "Another Sternberg 'Fish-Within-a-Fish' Discovery: First Report of *Ichthyodectes ctenodon* (Teleostei; Ichthyodectiformes) with Stomach Contents." *Transactions of the Kansas Academy of Science* 113: 197–205.

O'Shea, M., et al. 2013. " 'Fantastic Voyage': A Live Blindsnake (*Ramphotyphlops braminus*) Journeys Through the Gastrointestinal System of a Toad (*Duttaphrynus melanostictus*)." *Herpetology Notes* 6: 467–70.

Ritschel, C. 2019. "Great White Shark Chokes to Death on Sea Turtle." *Independent*, April 23, 2019. https://www.independent.co.uk/news/world/asia/great-white-shark-sea-turtle-choke-japan-fishing-a8882746.html.

Shimada, K. 2019. "A New Species and Biology of the Late Cretaceous 'Blunt-Snouted' Bony Fish, *Thryptodus* (Actinopterygii: Tselfatiiformes), from the United States." *Cretaceous Research* 101: 92–107.

Walker, M. 2006. "The Impossible Fossil: Revisited." *Transactions of the Kansas Academy of Science* 109: 87–96.

Cracking the Case of the Bone-Crushing Dogs

Van Valkenburgh, B., et al. 2003. "Pack Hunting in Miocene Borophagine Dogs: Evidence from Craniodental Morphology and Body Size." In *Vertebrate Fossils and Their Context: Contributions in Honor of Richard H. Tedford*, ed. L. J. Flynn, 147–62. *Bulletin of the American Museum of Natural History* 279. New York: American Museum of Natural History.

Wang, X., et al. 2008. *Dogs. Their Fossil Relatives & Evolutionary History*. New York: Columbia University Press.

Wang, X., et al. 2018. "First Bone-Cracking Dog Coprolites Provide New Insight Into Bone Consumption in Borophagus and Their Unique Ecological Niche." *eLife* 7, e34773.

To Catch a Killer

Wilson, J. A., et al. 2010. "Predation Upon Hatchling Dinosaurs by a New Snake from the Late Cretaceous of India." *PLOS One* 8, e1000322.

A Dinosaur-Eating Mammal

Hu, Y., et al. 2005. "Large Mesozoic Mammals Fed on Young Dinosaurs." *Nature* 433: 149–52.

Li, J., et al. 2001. "A New Family of Primitive Mammal from the Mesozoic of Western Liaoning, China." *Chinese Science Bulletin* 46: 782–85.

The Feeding Ground

Jennings, D. S., and S. T. Hasiotis. 2006. "Taphonomic Analysis of a Dinosaur Feeding Site Using Geographic Information Systems (GIS), Morrison Formation, Southern Bighorn Basin, Wyoming, USA." *Palaios* 21: 480–92.

Lippincott, J. 2015. *Wyoming's Dinosaur Discoveries.* Charleston, SC: Arcadia, 95.

Hell Pig Meat Cache

Benton, R. C., et al. 2015. *The White River Badlands: Geology and Paleontology*. Bloomington: Indiana University Press, 240.

Prothero, D. R. 2016. *The Princeton Field Guide to Prehistoric Mammals.* Princeton, NJ: Princeton University Press, 240.

Sundell, K. A. 1999. "Taphonomy of a Multiple *Poebrotherium* Kill Site—An *Archaeotherium* Meat Cache." *Journal of Vertebrate Paleontology* 19: 79A.

Prehistoric Dolls

Kriwet, J., et al. 2008. "First Direct Evidence of a Vertebrate Three-Level Trophic Chain in the Fossil Record." *Proceedings of the Royal Society B* 275: 181–86.

Smith, K. T., and A. Scanferla. 2016. "Fossil Snake Preserving Three Trophic Levels and Evidence for an Ontogenetic Dietary Shift." *Palaeobiodiversity and Palaeoenvironments* 96: 589–99.

UNUSUAL HAPPENINGS

Introduction

Arias-Robledo, G., et al. 2018. "The Toad Fly *Lucilia bufonivora*: Its Evolutionary Status and Molecular Identification." *Medical and Veterinary Entomology* 33: 131–39.

Brusca, R. C., and M. R. Gilligan. 1983. "Tongue Replacement in a Marine Fish (*Lutjanus guttatus*) by a Parasitic Isopod (Crustacea: Isopoda)." *Copeia* 3: 813–16.

Cressey, R., and C. Patterson. 1973. "Fossil Parasitic Copepods from a Lower Cretaceous Fish." *Science* 180: 1283–85.

Klompmaker, A. A., and G. A. Boxshall. 2015. "Fossil Crustaceans as Parasites and Hosts." *Advances in Parasitology* 90: 233–89.

Labandeira, C. C. 2002. "Paleobiology of Predators, Parasitoids, and Parasites: Death and Accommodation in the Fossil Record of Continental Invertebrates." *Paleontological Society Papers* 8: 211–49.

Welicky, R. L., et al. 2019. "Understanding Growth Relationships of African Cymothoid Fish Parasitic Isopods Using Specimens from Museum and Field Collections." *Parasites and Wildlife* 8: 182–87.

Parasite Rex

Brink, K. 2020. "Description of New Tooth Pathologies in *Tyrannosaurus rex*." *Society of Vertebrate Paleontology, 80th Annual Meeting, Abstracts*: 84.

Wolff, E. D. S., et al. 2009. "Common Avian Infection Plagued the Tyrant Dinosaurs." *PLOS One* 4, e7288.

Mass Mammal Strandings

Cordes, D. O. 1982. "The Causes of Whale Strandings." *New Zealand Veterinary Journal* 30: 21–24.

Häussermann, V., et al. 2017. "Largest Baleen Whale Mass Mortality During Strong El Niño Event Is Likely Related to Harmful Toxic Algal Bloom." *PeerJ* 5, e3123.

Pyenson, N. D., et al. 2014. "Repeated Mass Strandings of Miocene Marine Mammals from Atacama Region of Chile Point to Sudden Death at Sea." *Proceedings of the Royal Society B* 281: 20133316.

Soundly Sleeping Dragon

Gao, C., et al. 2012. "A Second Soundly Sleeping Dragon: New Anatomical Details of the Chinese Troodontid *Mei long* with Implications for Phylogeny and Taphonomy." *PLOS One* 7, e45203.

Rogers, C. S., et al. 2015. "The Chinese Pompeii? Death and Destruction of Dinosaurs in the Early Cretaceous of Lujiatun, NE China." *Palaeogeography, Palaeoclimatology, Palaeoecology* 427: 89–99.

Wang, Y., et al. 2016. "Stratigraphy, Correlation, Depositional Environments, and Cyclicity of the Early Cretaceous Yixian and ?Jurassic-Cretaceous Tuchengzi Formations in the Sihetun Area (NE China) Based on Three Continuous Cores." *Palaeogeography, Palaeoclimatology, Palaeoecology* 464: 110–33.

Xu, X., and M. A. Norell. 2004. "A New Troodontid Dinosaur from China with Avian-Like Sleeping Posture." *Nature* 431: 838–41.

Snap That!

Ballell, A. et al. 2019. "Convergence and Functional Evolution of Longirostry in Crocodylomorphs." *Palaeontology* 62: 867–87.

Bulstrode, C., et al. 1986. "What Happens to Wild Animals with Broken Bones?" *Lancet* 327: 29–31.

Eldredge, N., and S. M. Stanley. 1984. *Living Fossils*. New York: Springer, 291.

Hartstone-Rose, A., et al. 2015. "The Bacula of Rancho La Brea." *Science Series* 42: 53–63.

Huchzermeyer, F. W. 2003. *Crocodiles: Biology, Husbandry and Diseases*. Wallingford, UL: CABI, 352.

Irwin, S. 1996. "Survival of a Large *Crocodylus porosus* Despite Significant Lower Jaw Loss." *Memoirs of the Queensland Museum* 39: 338.

Marsell, R., and T. A. Einhorn. 2011. "The Biology of Fracture Healing." *Injury* 42: 551–55.
Philo, L. M., et al. 1990. "Fractured Mandible and Associated Oral Lesions in a Subsistence-Harvested Bowhead Whale (*Balaena mysticetus*)." *Journal of Wildlife Diseases* 26: 125–28.
Pierce, S. E., et al. 2017. "Virtual Reconstruction of the Endocranial Anatomy of the Early Jurassic Marine Crocodylomorph *Pelagosaurus typus* (Thalattosuchia)." *PeerJ* 5, e3225.
Stockley, P. 2012. "The Baculum." *Current Biology* 22: R1032–R1033.

Drama of a Drought

Brusatte, S. L., et al. 2015. "A New Species of *Metoposaurus* from the Late Triassic of Portugal and Comments on the Systematics and Biogeography of Metoposaurid Temnospondyls." *Journal of Vertebrate Paleontology* 35, e912988.
Colbert, E. H., and J. Imbrie. 1956. "Triassic Metoposaurid Amphibians." *Bulletin of the American Museum of Natural History* 110: 405–2.
Gee, B. M., et al. 2020. "Redescription of *Anaschisma* (Temnospondyli: Metoposauridae) from the Late Triassic of Wyoming and the Phylogeny of the Metoposauridae." *Journal of Systematic Palaeontology* 18: 233–58.
Lucas, S. G., et al. 2010. "Taphonomy of the Lamy Amphibian Quarry: A Late Triassic Bonebed in New Mexico, U.S.A." *Palaeogeography, Palaeoclimatology, Palaeoecology* 298: 388–98.
Lucas, S. G., et al. 2016. "Rotten Hill: A Late Triassic Bonebed in the Texas Panhandle, USA." *New Mexico Museum of Natural History and Science Bulletin* 72: 1–96.
Romer, A. S. 1939. "An Amphibian Graveyard." *Scientific Monthly* 49: 337–39.
Wells, K. D. 2007. *The Ecology and Behavior of Amphibians*. Chicago: University of Chicago Press, 1400.
Zeigler, K. E., et al. 2002. "Taphonomy of the Late Triassic Lamy Amphibian Quarry (Garita Creek Formation: Chinle Group), Central New Mexico." *New Mexico Museum of Natural History and Science Bulletin* 21: 279–83.

Eaten from the Inside Out

Eberhard, W. G., and M. O. Gonzaga. 2019. "Evidence that *Polysphincta*-Group Wasps (Hymenoptera: Ichneumonidae) Use Ecdysteroids to Manipulate the Web-Construction Behaviour of Their Spider Hosts." *Biological Journal of the Linnean Society* 127: 429–71.
van de Kamp, T., et al. 2018. "Parasitoid Biology Preserved in Mineralized Fossils." *Nature Communications* 9: 3325.

Dinosaur Tumors

Barbosa, F. H. d-S., et al. 2016. "Multiple Neoplasms in a Single Sauropod Dinosaur from the Upper Cretaceous of Brazil." *Cretaceous Research* 62: 13–17.

Bianconi, B., et al. 2013. "An Estimation of the Number of Cells in the Human Body." *Annals of Human Biology* 40: 463–71.

Dumbrava, M. D., et al. 2016. "A Dinosaurian Facial Deformity and the First Occurrence of Ameloblastoma in the Fossil Record." *Nature Scientific Reports* 6: 29271.

Ekhtiari, S., et al. 2020. "First Case of Osteosarcoma in a Dinosaur: A Multimodal Diagnosis." *Lancet* 21: 1021–22.

Gonzalez, R., et al. 2017. "Multiple Paleopathologies in the Dinosaur *Bonitasaura salgadoi* (Sauropoda: Titanosauria) from the Upper Cretaceous of Patagonia, Argentina." *Cretaceous Research* 79: 159–70.

Hao, B.-Q., et al. 2018. "Femoral Osteopathy in *Gigantspinosaurus sichuanensis* (Dinosauria: Stegosauria) from the Late Jurassic of Sichuan Basin, Southwestern China." *Historical Biology* 32: 1028–35.

Masthan, K. M. K., et al. 2015. "Ameloblastoma." *Journal of Pharmacy and Bio Allied Sciences* 7: 276–78.

Macmillan Cancer Support. 2020. "What Is Cancer?" https://www.macmillan.org.uk/cancer-information-and-support/worried-about-cancer/what-is-cancer.

Rothschild, B. M., et al. 1998. "Mesozoic Neoplasia: Origins of Haemangioma in the Jurassic Age." *Lancet* 351: 1862.

Rothschild, B. M., et al. 1999. "Metastatic Cancer in the Jurassic." *Lancet* 354: 398.

Rothschild, B. M., et al. 2003. "Epidemiologic Study of Tumors in Dinosaurs." *Naturwissenschaften* 90: 495–500.

Fossilized "Farts"

Penney, D. 2016. *Amber Palaeobiology. Research Trends and Perspectives for the 21st Century.* Manchester, UK: Siri Scientific, 127.

Poinar, G. O., Jr. 2009. "Description of an Early Cretaceous Termite (Isoptera: Kalotermitidae) and Its Associated Intestinal Protozoa, with Comments on Their Co-Evolution." *Parasites & Vectors* 18: 1–17.

Poinar, G. O., Jr. 2010. "Fossil Flatus: Indirect Evidence of Intestinal Microbes." In *Fossil Behavior Compendium*, ed. A. J. Boucot and G. O. Poinar Jr., 22–25. Boca Raton, FL: CRC.

Rabaiotti, D., and N. Caruso. 2017. "Does It Fart?" London: Quercus, 144.

Could It Be Dinosaur Pee?

Black, R. 2014. "Laelaps. The Surprising Science of Dinosaur Pee." *National Geographic*, February 12, 2014. https://www.nationalgeographic.com/science/phenomena/2014/02/12/the-surprising-science-of-dinosaur-pee/.

Fernandes, M., et al. 2004. "Occurrence of Urolites Related to Dinosaurs in the Lower Cretaceous of the Botucatu Formation, Paraná Basin, São Paulo State, Brazil." *Revista Brasileira de Paleontologia* 7: 263–68.

Martin, A. J. 2014. *Dinosaurs Without Bones. Dinosaur Lives Revealed by Their Trace Fossils.* New York: Pegasus, 460.

McCarville, K., and G. Bishop. 2002. "To Pee or Not to Pee: Evidence for Liquid Urination in Sauropod Dinosaurs." *Journal of Vertebrate Paleontology* 22: 85 A.

Souto, P. R. F., and M. A. Fernandes. 2015. "Fossilized Excreta Associated to Dinosaurs in Brazil." *Journal of South American Earth Sciences* 57: 32–38.

Wedel, M. 2016. "Yes, Folks, Birds and Crocs Can Pee." Sauropod Vertebra Picture of the Week, January 28, 2016. https://svpow.com/2016/01/28/yes-folks-birds-and-crocs-can-pee/.

Index